HEALTHCARE ROBOTS

Emerging Technologies, Ethics and International Affairs

Series editors:

Jai C. Galliott, The University of New South Wales, Australia
Avery Plaw, University of Massachusetts, USA
Katina Michael, University of Wollongong, Australia

This series examines the crucial ethical, legal and public policy questions arising from or exacerbated by the design, development and eventual adoption of new technologies across all related fields, from education and engineering to medicine and military affairs.

The books revolve around two key themes:

- Moral issues in research, engineering and design.
- Ethical, legal and political/policy issues in the use and regulation of Technology.

This series encourages submission of cutting-edge research monographs and edited collections with a particular focus on forward-looking ideas concerning innovative or as yet undeveloped technologies. Whilst there is an expectation that authors will be well grounded in philosophy, law or political science, consideration will be given to future-orientated works that cross these disciplinary boundaries. The interdisciplinary nature of the series editorial team offers the best possible examination of works that address the 'ethical, legal and social' implications of emerging technologies.

Forthcoming titles:

Legitimacy and Drones
UCAVs for Cross-Border Counterterrorism
Edited by Steven J. Barela

Commercial Space Exploration
Ethics, Policy and Governance
Edited by Jai Galliott

Healthcare Robots

Ethics, Design and Implementation

AIMEE VAN WYNSBERGHE
University of Twente, the Netherlands

ASHGATE

Published by
Ashgate Publishing Limited
Wey Court East
Union Road
Farnham
Surrey, GU9 7PT
England

Ashgate Publishing Company
110 Cherry Street
Suite 3-1
Burlington, VT 05401-3818
USA

www.ashgate.com

British Library Cataloguing in Publication Data
A catalogue record for this book is available from the British Library.

The Library of Congress has cataloged the printed edition as follows:
Wynsberghe, Aimee van, author.
Healthcare robots : ethics, design and implementation / by Aimee van Wynsberghe.
p. ; cm. -- (Emerging technologies, ethics and international affairs)
Includes bibliographical references and index.
ISBN 978-1-4724-4433-2 (hardback) -- ISBN 978-1-4724-4434-9 (ebook) -- ISBN 978-1-4724-4435-6 (epub)
I. Title. II. Series: Emerging technologies, ethics and international affairs.
[DNLM: 1. Patient Care--ethics. 2. Robotics--ethics. 3. Equipment Design. 4. Robotics--instrumentation. 5. Social Values. QT 36.2]
R857.R63
610.28'4--dc23

2014046618

ISBN 9781472444332 (hbk)
ISBN 9781472444349 (ebk – PDF)
ISBN 9781472444356 (ebk – ePUB)

MIX
Paper from responsible sources
FSC www.fsc.org FSC® C013985

Printed in the United Kingdom by Henry Ling Limited, at the Dorset Press, Dorchester, DT1 1HD

Contents

For Aria and Scott

List of Tables

Preface

There are many who believe in the wonderful potential of robotics technology and there are many who are terrified of enveloping robots into society. Both groups insist that we should proceed with caution when it comes to the development and implementation of robots to avoid negative consequences. To my great astonishment, there are also individuals who believe there are no ethical concerns to consider in the design of care robots because none have revealed themselves so far.

In response to this intuition I suggest one look at any new technology in healthcare and to reflect on the unanticipated ethical considerations: new technologies shape cultural norms and assumptions about not only what is healthy but what is the appropriate standard of care. Once robots enter the picture, healthy individuals may be defined by their lack of reliance on a personal robot in the home. Vice versa, you could be considered "unhealthy" if you need a robot to assist with some part of your activities of daily living. From this we might consider if there will be repercussions for insurance companies? Care robots integrated into hospital contexts may become standard practice once proven to provide effective care. It may be possible one day that healthcare institutes are obliged to provide care with the use of robots. This is not far-fetched when you consider the use of new surgical tools, such as endoscopic tools, that faced stern criticism when first introduced but which are now standard practice in most surgical suites.

But these are not even the most interesting ethical considerations! As a robot ethicist, what I find to be the most pressing questions are those that deal with how robots are thought to work in care contexts and how this relies on a variety of assumptions about what care looks like in context. From these assumptions, both implicit and explicit, the robot is designed. Through this design, assumptions about values and norms are reinforced; practices are reinforced or are changed. It is this cycle that requires careful reflection in order for us as a society to decide when and where we want to have these robots.

In order to do justice to a book that dealt with understanding care in the context of a hospital and/or nursing home I spent time observing and volunteering in those contexts. It was during those moments that I learned of the wonderful benefits that robots could provide in a healthcare system but it was also during that time that I understood the very real threats to care that robots posed. This was one of the factors that shaped my opinion on the design and use of care robots, namely to take a middle ground and insist on the ethical shaping of the technology rather than on stopping the technology altogether.

This book is meant to pause and take a moment to consider all the good and bad attributed to robots in healthcare, and to bring these considerations into the design process. This is not the central aim of the book; rather, it is the starting point. With this book I want to take a proactive stance to robot development and to show that it is possible to shape the coming care robots in a way that takes the values in care as a central focus: care robots designed based on care values. These values are not abstract unattainable values but are in fact definable and attainable once we have a very specific description of a care practice. So my task here is to bring together the perspectives of multiple disciplines in order to join them in a framework for care robot design. The goal of my future career will be to put this framework into practice in the hopes of integrating ethics directly into the industrial context where care robots are designed and developed. So, here's to the future!

Aimee van Wynsberghe

Acknowledgements

A huge thank you to my mentor Professor Dr Philip Brey at the University of Twente. A warm thank you to my fellow robot colleagues, professors Noel Sharkey and Wendell Wallach, for your inspiration, feedback and support. Most notably thank you to Professor Joan Tronto for your insights into care that helped guide my thinking.

Thank you to my friends and family in Canada and the Netherlands for your encouragement and love. Most importantly, thank you to my amazing husband Scott and to my incredible daughter Aria. I am forever in awe of you both.

List of Abbreviations

AC	Affective Computing
ADLs	Activities of Daily Living
AI	Artificial Intelligence
ANT	Actor-Network Theory
CC	Care-Centred
CCVSD	Care-Centred Value-Sensitive Design
HAL	Hybrid Assistive Limb
HFR	Human-Friendly Robots
HULC™	Human Universal Load Carrier
ISO	International Organization for Standardization
RIBA	Robot for Interactive Body Assistance
STS	Science and Technology Studies
VSD	Value-Sensitive Design
WHO	World Health Organization

Introduction

A Revolution in Healthcare

Welcome to a revolution in healthcare. Increasingly, policy makers and healthcare providers are turning their attention to robots as one solution among others to overcome shortages in healthcare resources and personnel anticipated worldwide. Interaction with robotic pets, such as Sony's AIBO or the robot seal Paro, are shown to have positive physiological benefits on elderly people. Service robots, such as Aethon's TUG® robot or the HelpMate™, are currently used in hospitals across the United States for the delivery of sheets and medications. With the widespread introduction of robots used in healthcare, the "robot revolution" has spawned what can only be referred to as a revolution in healthcare. This book addresses the initiative to create and use care robots and the many ethical questions surrounding their design and implementation.

As we come into the twenty-first century, the ageing population is already a major demographic worldwide and will continue to increase dramatically. The care-givers available to care for this large segment of the population are woefully outnumbered by this "boomer" generation. According to the World Health Organization (WHO), while life expectancy is increasing, fertility rates are declining around the world (WHO, 2010). The continued anticipated increase in this population group is reason for concern, as is the challenge of providing care for the population is general. This will be hampered by a lack of resources, a competition for healthcare services, shortages of personnel and care providers, and a changing pattern of need. It will be a test for healthcare systems around the world to maintain standards of care let alone improve quality of care. Policy makers are grappling with the question of how such setbacks are to be mitigated. One possible answer: robots.

Although the influence of popular culture conjures images of human-like robots, such as Star Wars' C-3PO, performing a surgery on a human, this not the case. Currently, in healthcare applications, robots are now available to help in surgical tasks that a surgeon couldn't otherwise complete with the same precision. Such robots are big and bulky, machine-like in appearance and require the direct input of a human user in order to execute an action. Hospitals and healthcare facilities are using robots in rehabilitation treatments, the sorting of medications, delivery of food, and as a communication platform between patients and physicians when geographical boundaries separate the two. These robots are already commercially available and used in hospitals in the US, Canada, Europe, and Japan.

In elderly care facilities in Japan, robot teddy bears monitor and assess the functioning of patients and report back to staff (Sharkey and Sharkey, 2011). The PaPeRo robot is used in a similar way for the care or monitoring of children (ibid.). Such trends are expected to continue in order to facilitate remote monitoring of patients and are even thought to be used in the event that patients are quarantined.

Outside these examples, the latest developments in healthcare robotics are those intended to assist the nurse in his/her daily tasks. These robots, now referred to as care robots, may be used by the care-giver and/or the care-receiver for a variety of tasks from bathing, lifting, feeding, to playing games. They may have any range of robot capabilities and features and may be used in a variety of contexts, i.e. nursing home, hospital or home setting. They are integrated into the therapeutic relationship between care-giver and care-receiver with the aim to meet a range of care needs of users. With this in mind, these robots are expected, and hoped, to overcome the lack of healthcare personnel and resources in large institutes or in specific instances to allow persons to stay in their home without having to live in a care institute (as in the care of elderly or rehabilitative persons).

To be clear, in this book I am not making the claim that these robots should be made and used for all care activities. I am also not claiming that these robots should never be made or used in care practices. My objective is to take a middle ground between these two extremes – to find the limits within which care robots should be designed and used. This opinion is grounded in the belief that there are benefits a care robot can provide but there are also a number of ethical problems that may arise with the use of a care robot.

In terms of the first point – that care robots can provide a benefit – a care robot presents the option of providing impartial care 24/7. It cannot be denied that care is required 24/7 and it is not possible for one care provider alone to provide this kind of assistance, in a hospital or home setting. With respect to impartial care, it cannot be denied that not all patients receive the same kind of care; some care receivers will have a better rapport or relationship with one care giver over another and may receive better care because of it. Moreover, in practice many nurses in the hospital feel an affinity for some patients over others, especially when patients themselves are abusive. In summary, each patient is treated differently for a variety of reasons.

Research has shown that in a homecare setting there is the risk of maltreatment of patients: "a relationship with a care recipient can evoke a multitude of attitudes and behaviours. At times, deplorable traits can emerge.

In fact, individuals suffering from debilitating illnesses such as dementia are sometimes mistreated by family members" (Cooper, 2009, cited in Borenstein, 2011, p. 257).

The fear of such treatment is not exclusive to a home setting: "at the present moment when the costliness of labour-intensive care is foremost in the minds of citizens" (Razavi, 2007), we frequently hear about abusive or inadequate forms of care (Tronto, 2010, p. 163). In the nursing home patients are often reported to be abused physically, emotionally or psychologically (Pillemer, 1990; Payne, 1995; Podnieks, 1990).

Seen through this lens, a robot in place of a nurse for certain tasks or at certain times in the night/day presents the potential to overcome concerns of impartiality and abuse as well as providing care at all times of the day: the robot may be used as a way of regulating the behaviour of human care-workers to avoid any risk of patient abuse or maltreatment.

But care robots are not only thought to provide a benefit for preventing patient abuse, they are thought to provide a real benefit for assisting care givers. Care givers face a number of physical and emotional challenges in their day-to-day work that a robot could help with, such as lifting or bathing patients. But aside from these better known examples are a host of examples where the day-to-day activities of a nurse require that they risk their own safety or wellbeing. I will discuss examples of these later on in this book.

The picture of robots being such a help to society and relieving us of tiresome or mundane burdens is not always the picture presented in popular culture discourse like books and movies. In movies, Western societies are presented with a multitude of dystopian futures in which humans become lazy and completely dependent on robots for their well-being (Morris, 2008) – consider the Pixar movie WALL-E – or humans suffering at the hands of robots when the robots override the decisions of humans – consider the movie *2001: A Space Odyssey* (Kubrick, 1968). Although entertaining, literature and movies fuel the views and beliefs of popular discourse and have left some societies in fear of living with robots.

The image of robots being detrimental to society is not only in the movies and literature; robot ethicists and potential users don't necessarily see the introduction of robots into care as being a good thing. Care robots, in particular, pose certain ethical concerns specific to the tradition of care. Scholars have written about a variety of ethical concerns including: the potential for social isolation when a robot is used in place of a human for social and emotional caring tasks (Sparrow, 2006); the potential to threaten the rights of elderly persons (Sharkey and Sharkey, 2012; Sharkey and Sharkey, 2011); minimizing the opportunity for (self) growth of care-givers and other users (Vallor, 2011); a potential to threaten the privacy, security and confidentiality of a patient when the robot is used to communicate information from one setting to another; safety concerns in terms of the physical well-being of patients; the potential to threaten the quality of care of patients (Coeckelbergh, 2010); risks of ageist discrimination; concerns of distributive justice or health equity. For the last point, the question concerns whether or not developing countries will have access to the technology or, whether the use of robots will be directed towards those who lack a certain social status (robots used for the care of prisoners, elderly persons, children, handicapped persons, etc.).

All these issues point out the very real and pressing ethical considerations related to care robots for which there are currently no (adequate) answers. A large portion of the difficulty in ethically assessing the design, development and use of care robots has to do with knowing what questions to ask. Should the ethical evaluation of care robots address the initiative to use such robots and the assumptions leading to such an initiative? An ethical evaluation of care robots could focus on how their introduction will impact the organization and provision of care. Perhaps the most appropriate course of action would be to ethically steer the design and development, or, perhaps the ethics of care robots ought to centre on the implementation of the robot?

Each of these questions targets a different point in the design process of a care robot thereby appealing to a different set of stakeholders (designers vs. users) or a different context (the lab vs. the hospital/home setting). Given that care robots will have an ethical impact on a range of stakeholders, in a range of contexts, it follows that their evaluation ought to encompass all of the aforementioned questions. At the centre of any analysis of a care robot ought to be a reflection on the impact the robot will have on the provision of good care.

Ethics and Care

If we take the starting point to be that the initiative to create and use care robots rests on the belief that care robots will maintain a high standard of care, or perhaps even improve care, then the main ethical focus needs to be how care is understood; what is care, what is good care, and how is this achieved and/or evaluated?

At the root of all the ethical issues addressed to date appears to be an ambiguity of what care is, how it is structured, what it involves and what it means. This only adds to the problem of articulating when care is good, for whom and what elements make it good. As psychoanalyst Sherry Turkle eloquently points out in her book *Alone Together* (2011), with the current generation of robots, we as a society are afforded the opportunity to reflect on the values of societal importance and to safeguard their place or alternatively allow for a trade-off between values.

This opportunity is what Turkle refers to as "the robotic moment" and is the situation we are currently in. But this is more than an opportunity claims Turkle, it is a necessity. Care robots offer us the opportunity to reflect on care – what it is, how it is achieved. For authors like Shannon Vallor, this reflection involves paying special attention to the goods at stake for the care-giver when a robot is used (Vallor, 2011). For Sparrow and Sparrow this involves recognition of the significance of the component of human presence in care (Sparrow, 2006). For Sharkey and Sharkey, this involves recognition of the rights of vulnerable demographics and how a care robot may impact such rights (Sharkey and Sharkey, 2012).

I would like to go even further than this and examine the very root(s) of care. Such a feat demands an understanding of care conceptually, in terms of the many values that comprise care, as well as understanding care in context, in terms of the actions and interactions between care providers and care-receivers that realize care values.

In the care ethics tradition, care is described as a process with many actors ranging from nurses, physicians, and pharmacists to secretaries, cleaning staff, technicians and beyond. The process involves the many care activities for a patient but also for the care institute. The many actions in care that make up the overall process of care are referred to as care practices. Consequently, care as a concept is distinguished from contextualized care. The former I refer to as "care": the conceptual dimension of care that centres on a valuation of another, concepts like human dignity and a relationship with the good life. The latter, contextualized care, I refer to as care practices, which provide meaning to abstract values such as human dignity. Understanding "care practices" is necessary to give meaning to "care" and vice versa.

To do just that we need a framework for understanding a care practice; how care values are made real, how roles and responsibilities are distributed, and how meanings are established. Only by understanding these variables can we come to understand: the role the care robot will play once introduced; the responsibility and meaning the robot will have; whether or not the robot preserves the expression of values or alters them, and if so in what way. When we understand what is happening in care at the contextualized level (what can also be referred to as the micro level), we may begin to understand the significance of the robot at the same level.

With this in mind, the goal of this book is to understand how care robots, used in care practices, can be designed and implemented in a way that supports and promotes the fundamental values in care. This central question takes into consideration all of the aforementioned questions, namely; how will the care robot impact the expression of care values, how will the care robot impact the distribution of roles and responsibilities, and what meaning will the care robot take on? To facilitate this kind of ethical evaluation of care robots I will create a framework for understanding the web within which the care robot will enter and the potential impact the robot might have. Of equal importance is not only evaluating current care robot prototypes but how to design future care robots in an ethically sound way. The framework will not only allow for this evaluation but will target issues of future design as well.

A Framework for Evaluating Care Robots

How will such a framework be created, how will it be used and what is its purpose? Chapter 1, "Designing Robots with Care", explains in detail how I have chosen to address these questions. The framework is both conceptual, in that it allows for an understanding of how values are manifest in care practices among actors (human and non-human), and normative in that it allows for the analysis and evaluation of the impact a robot may have on the promotion and expression of care values in context. In my work, I draw upon a number of theoretical approaches and methodologies, and this chapter aims to explore many of these approaches and concepts. In particular, I draw on the concept of embedded values (Nissenbaum, 2001), Value-Sensitive Design (VSD) (Friedman et al., 2006), and the care ethics tradition, namely the work of care ethicist Joan Tronto (1993; 2010).

In order to address the relationship between a care robot and contextualized care, we must first understand what care values are and how they come into being. Chapter 2, "Understanding Care in Context", goes into a conceptual analysis of the dominant values of the care ethics tradition. Special attention is paid here to the description of a care practice and the significance of understanding care tasks as practices rather than as tasks. I explore the fundamental values in care from a top down approach beginning with the abstract values articulated by the World Health Organization and how they become concrete when understood in context. This chapter reveals three important findings: 1) values are manifest (or co-produced) through the actions and interactions among actors (human and non-human) in a network for a particular practice in a specific context; 2) a care practice is a small piece in the holistic vision of care as a process (Tronto, 1993); and 3) the therapeutic relationship is the vehicle for the manifestation of care values.

Chapter 3, "Robots and Robot Capabilities", deals with the definition of a robot. In this chapter I present the variety of robot capabilities and features. Chapter 4 addresses the definition of a care robot. In addition, I present existing care robot prototypes currently in use or still in the developmental stages. This chapter reveals the impossibility of translating human capabilities into robot capabilities independent of contextual variables (the care practice and the actors involved). From this I conclude that without an understanding of the context within which the care robot will be applied or the practice for which it will be used, one is not capable of truly understanding the impact the robot may have. Consequently, I begin to set the stage for the various components of the framework, namely, that context and practice must be made explicit if one is to understand the impact the care robot will have.

Here we are faced with the question: how will all of this be used? Chapter 5, "A Framework for Evaluating the Design of Care Robots", outlines and describes the components of the framework and the justification for their place within the framework. I refer to the framework as the Care-Centred (CC) framework given the focal role the care perspective plays in its creation and usage. This framework is then used for two types of value-based analyses: 1) for retrospective evaluations of current care robots; and 2) in the prospective design of future care robots.

Chapter 6, "Care Robots and the Practice of Lifting", investigates both the practice of lifting, and the current robots delegated for such a practice. Two care robot designs used for the lifting of patients are compared with each other to illustrate how differing robot capabilities result in divergent visions of the resulting care practice. Each robot is examined using the current practice of lifting to understand the way in which a care robot might be used to re-integrate values lost in the first wave of automation (i.e. the mechanical lift for lifting) as well as how the robot may impede the promotion of necessary values. The aim of this chapter is to make clear the relationship between the technical capabilities of the robot and its impact on the resulting care practice.

Chapter 7, "The Future Design of Care Robots: The Care-Centred Value-Sensitive Design Approach", outlines the overall approach referred to Care-Centred Value-Sensitive Design (CCVSD). The approach begins at the moment of idea generation and continues through various design prototypes. This is illustrated by presenting two novel care robots: the *wee-bot* and the *roaming toilet*. This chapter sketches a vision of the role of the ethicist as a member of the design team engaged in a didactic process throughout the design process. This chapter also investigates the moral impact of a care robot in terms of moral agency. Although the moral agency has been addressed implicitly throughout the preceding chapters, my aim in this chapter is to explicitly discuss the moral status of robots and the consequences such a discussion has on the design of future care robots.

When we take into consideration the idea that the robot is being designed according to a specific use, one that acts to promote care values and one that determines a particular distribution of roles and responsibilities, we must also consider how the care robot will be introduced, or rather, how the care robot *ought* to be introduced. Thus, the CCVSD approach does not end with the resulting artefact. It must also address the implementation of the care robot to ensure that the values embedded within the robot are in fact realized in their intended use context. This is the goal of Chapter 8, "Conclusion: Implementing Care Robots with Care".

You will notice in this book that I leave out the issue of stakeholder involvement or user preferences (i.e. participatory design). This is not because I believe the user doesn't have a say or shouldn't be asked about their preferences but rather because I believe we cannot derive normative conditions from user preferences. We can, however, derive normative standards from the conceptual analysis of ethics literature that explicitly addresses such issues (e.g. the care ethics tradition). There are limits to what an institution can provide but the hope is that by taking the ideal of care as the normative goal we can come as close as possible to *good* care.

With the CCVSD approach my goal is to foster an interdisciplinary approach and a division in moral labour, throughout the design, development, and implementation of care robots. Given the initiative to bridge disciplines, this book is intended to be read by individuals interested in the topic of care and the introduction of care robots from a variety of fields e.g. design studies, applied ethics, robot ethics, computer science, computer ethics and so on. As such, each field of study that I draw upon is presented in the most straightforward manner possible.

The creation of the CCVSD approach is meant to mark the "robotic moment", coined by Sherry Turkle. This "robotic moment" that Turkle speaks of demands that care robots undergo meticulous ethical evaluation. This "robotic moment" also demands that our traditional conceptions of relationships, of the meaning of care and of what it means to be human are questioned and subject to re-interpretation. My response to the claim of Turkle is to structure both this revolutionary technology in healthcare applications as well as structuring healthcare institutions in a way that supports the introduction of the robot and supports the roles, responsibilities and valuation of healthcare workers. Thus, not only can one consider the technology of care robots as a revolution in healthcare, but designing and implementing them according to the CCVSD approach is also a revolution; one that re-affirms and supports the values of the healthcare tradition along with the roles of healthcare providers.

Chapter 1
Designing Care Robots with Care

Introduction

Imagine an elderly woman – let's call her Margaret – lives alone in her home after the death of her beloved spouse. Margaret would prefer to remain in the comfortable and familiar surroundings of her own home for as long as possible to continue with some independence and dignity. Margaret requires monitoring as her health is declining with age. To avoid moving to a nursing home or other care facility, Margaret has purchased a home-care robot that can: remind her to take her medications; fetch items for her if she is too tired or is already in bed; help with simple cleaning tasks; and, can facilitate her staying in contact with her family, friends and healthcare provider via video chat. The robot holds the promise of independence for Margaret while maintaining a high level of care in the comfort of her own home.

The above scenario depicts the potential of home-care robots for one particular demographic and although it is not technologically feasible at this moment in time, the hope is that it will be realized in the not too distant future. As we come into the twenty-first century, the challenges of providing care for the elderly as well as the population in general will be constrained by: a lack of care resources, competition for healthcare services, and shortages of care personnel. Increasingly, policy makers and healthcare providers are turning their attention to robots as one solution among others, to meet the needs of patients in light of these foreseen challenges.

These robots, called care robots, are currently being designed and developed to be integrated into morally charged contexts like a hospital, nursing home or home-care setting. Based on the contexts within which they will be placed and the roles they will be assigned, roboticists claim that robots ought to be endowed with moral reasoning capabilities. In other words, the robot can decide what to do based on some conception of good/bad, right/wrong. This claim is incredibly problematic when we consider the relationship between moral agency and moral responsibility: a moral agent must bear moral responsibility for the consequences of his/her actions.

The issue of responsibility is of the utmost importance in healthcare contexts and in the therapeutic relationship: a human care-giver must be morally responsible for the outcome of care actions. The professionalization of medicine and nursing is grounded on this fact. This begs the question whether or not a robot can be morally responsible for the outcome of its actions and also whether or not we can consider a robot to be a moral agent. One of the things I will argue in this book is that the robot is not a moral agent but a moral factor given its impact on the moral decision making of the human care givers. This argument has significant repercussions for the design of robots in care contexts: they should be intentionally designed to avoid roles which require moral responsibility. This is only one of the many reasons why it is imperative that care robots undergo rigorous ethical reflection and evaluation.

Evaluating care robots is complicated for a variety of reasons; we must examine the question of "how" to evaluate (which ethical theory to apply or indeed if there is one theory that is sufficient), as well as the question of "what" to evaluate (the initiative to use care robots, their design, or their introduction) and, at the same time, we must examine and untangle the ethically good from the ethically bad uses.

The introduction of this work gave an overview of how care robots are seen to be beneficial in care as well as how their use raises ethical concerns. Accordingly, the question to ask is not whether or not we should make them, but *how* they should be made, and what they ought to be used for. I do not deny the development of this technology nor do I support the blind acceptance and use of this technology; rather, I am seeking a way in which the technology can be designed and made so that it can support widely held cultural care values.

In this book I suggest a way in which care robots can be ethically evaluated in a purposeful manner through an analysis of the relationship between robot design and the realization of care values through the use of the robot. Perhaps more importantly, I am also suggesting and arguing in favour of a more proactive stance: that care robots ought to be designed in a way that ensures the core care values are integrated in the design process and embedded into their design (i.e. their technical content). In this way it is hoped that the care robot is used in a way that realizes care values.

My suggestion for designing care robots *with* care relies on a normative approach to design that I call the Care-Centred Value-Sensitive Design Approach (CCVSD). I emphasize *with* care as a way of indicating that these care robots will be, and ought to be, designed in a manner that acknowledges the impact they will have on the provision of good care. Moreover, designing *with* care also refers to the fact that designers are intentionally working to embed care values into a robot's technical content.

This book is dedicated to presenting the theoretical justification for the CCVSD approach as well as demonstrating its utility. I aim to show how the approach can be used in a retrospective manner – to evaluate current care robots – and how the approach can be used prospectively – to steer the design of future care robots from the moment of idea generation up to the moment of implementation. But first, let us have a look at the technology we are discussing.

What is a Care Robot?

There is much confusion surrounding robots in terms of how they are defined and what they are currently capable of. This is due, in part, to the technical knowledge required to understand their functioning but also due to the role the media has played in shaping the image of a robot in the minds of society. The image given by the media, represented by Star Wars' charming C-3PO, Star Trek's endearing Data, and Pixar's adorable WALL-E all represent a class of robots not yet realized by today's technology. Current applications of robots are not capable of interacting in this social way (charming, endearing or adorable) as depicted in the media. Researchers are, however, deeply engrossed in the pursuit of creating robots that will one day communicate in this way.

In terms of defining care robots, there is not one capability, appearance, or function that is exclusive to a care robot. Care robots may be used in the home, hospital, nursing home or other setting. They may be used to assist in, support, or provide care for the elderly or otherwise vulnerable persons by providing assistance in care-giving tasks, monitoring a patient's health status and/or providing social care or companionship (Sharkey and Sharkey, 2012). Care robots may have any number and range of capabilities from planar locomotion (vs. stationary) to voice, facial or emotion recognition. They may appear machine-like, animal- or creature-like, or human/humanoid-like. Additionally, they may have varying degrees of autonomy – the amount of human involvement required for the robot to complete its tasks.

Today's commercially available healthcare robots include surgical robots (e.g. the daVinci® surgical system), delivery robots (e.g. the TUG® and HelpMate™ robots) and the Paro robot for social interactions. Robots in the testing phases include: rehabilitation robots for stroke survivors, robots to assist with lifting (e.g. the RI-MAN or RIBA [Robot for Interactive Body Assistance] robot to pick up patients and move them from one place to another), robots for bathing disabled patients (e.g. the bathing cabin robot for automatic washing and rinsing used in the Horsens Kommune), and robots for feeding patients (e.g. the Secom MySpoon robot) among others. For the roboticist, there is no limit to what a robot will be capable of in the future.

What connects all these devices is that they are machines capable of carrying out a complex series of actions automatically, usually programmed by computer scientists and engineers (i.e. roboticists). Thus, they have physical embodiment (they are not virtual characters), can act in their environment based on information they have sensed, and they have some range of automation. Although the development of robots used in a healthcare context is still in its infancy there are big plans for the future. Make no mistake, the robots are coming! The question then is: what will this new technology do to the age-old practice of care-giving?

Creating a Framework for the Ethical Evaluation of Care Robots: The Concept of Embedded Values

In this book I presuppose that ethics ought to be incorporated into the design process of robots (van Wynsberghe, 2013a; van Wynsberghe and Robbins, 2014; van Wynsberghe, 2014). This is in contrast to the current work of ethicists and robot scholars who address the ethical concerns retrospectively: once the robot has been made and has a clear application/use (Sharkey and Sharkey, 2011; Sparrow and Sparrow, 2006). The limitation of such an analysis is that the impact of ethics is limited to the implementation exclusively, and does not include any ethical impact from the resulting design.

Contemporary computer ethics and ethics of technology teach us that there are values embedded into the design of technologies (Friedman et al., 2006; Nissenbaum, 2011; Brey, 2014). This is known as the embedded values concept. An embedded value refers to value as the consequence of using a technology or artefact. If we take a value to be something good, something desirable, something that we want to have happen, then a value in the embedded sense refers to a good consequence of using a technology. As an example, the value of privacy is a consequence of using the phone capabilities (which do not allow for the tracking or tracing of phone conversations) provided by the company Silent Circle.

Of course one must acknowledge that a technology can promote a value while at the same time limit (or prohibit) the promotion of another competing value. Added to this a technology can have more than one use (i.e. dual-use). Take the example of Silent Circle given above. The technology is intended to be used by those in dangerous countries or abusive situations requiring privacy and protection. Although this is the intended use it is still possible that the technology might conceivably be used by unintended users for deplorable uses.

The argument remains that values are manifest during and through, the use of a technology. From this idea certain researchers have concluded that if a technology can realize a value (i.e. can bring a value into existence) then we should be able to intentionally design technologies to realize specified values of ethical importance. This idea led to the approach known as Value-Sensitive Design (VSD) (Friedman et al., 2006). Value Sensitive Design is a computer ethics approach dedicated to systematically incorporating a list of 12 ethical values into computer systems.

Value Sensitive Design has been praised for its ability to account for values in the design of computer systems as well as other technologies. It is the first approach that targets the design process as the place for value analysis and demands that the relationship between the design/use of a technology and the realization of values be considered. The approach consists of a tri-partite methodology to include conceptual, technical and empirical investigations. Conceptual analysis refers to an understanding of the value constructs in a philosophical sense as well as a practical one, i.e. in the context of use. Technical investigations refer to how the technology will work in context and the relationship with values. Empirical investigations involve questioning stakeholders about their preferences and behaviours with respect to the context in which the technology will be used. There are a range of methods for eliciting stakeholder involvement and uncovering the values of importance (e.g. value scenarios, envisioning workshop). There are also methods for identifying the values that cannot be limited or the ones considered the most important to stakeholders. These are referred to as value dams and flows respectively. These investigations are iterative rather than linear.

The approach I develop here, the CCVSD approach, relies on the starting point and basic skeleton of VSD but with notable differences. For starters, I am not creating a robot and thus my technical investigation will not involve actual experiments of users with the technology in a context of use. I do, however, deal with the technical content of the robot in great descriptive detail. Second, as I mentioned in the introduction I do not embark on stakeholder analysis to gather user preferences. Instead, the CCVSD approach generates a set of normative standards to follow rather than outlining the range of user preferences for a technology.

The CCVSD approach also shares the iterative aspect of VSD. In practice, the conceptual investigation of values in context alongside their embedding into a technology are overlapping practices that cannot be separated. Relating components to one another, as I will do, only strengthens the fluidity and consistency of the approach.

As with any new approach, alongside the strength and popularity of VSD come various criticisms. The criticisms centre on: the lack of a normative foundation for ethical evaluation (Manders-Huits, 2011); the lack of clarity surrounding the concept of value and/or embedded value (Van de Poel, 2009); the lack of understanding of how values are translated into design requirements (Van de Poel and Kroes, 2014); and, the disconnect between the intended values of engineers and the values realized once the artefact is used in context (ibid.). I will address each of these criticisms individually in considerable detail throughout the chapters of this book.

With respect to the first criticism, the lack of a normative foundation, the fear is that VSD relies on intuition rather than on an ethical tradition to ground it normatively. To mitigate this concern I use the care ethics tradition to provide the normative foundation for making any claim with respect to the robot's design. For this reason, among others, I refer to the approach developed here as Care Centred – essentially placing care ethics as a focal point in the approach.

Designing Care Robots for Care

To give a brief overview of the CCVSD approach, it consists of a framework of components of ethical significance (see Table 1.1) along with a "user manual" for evaluations. The framework, what I refer to as the Care-Centred framework, is a list of components to take into consideration in the evaluation of a care robot: the context of use, the care practice, the actors involved, the type of care robot (its capabilities, appearance etc.) and the list of values involved for the described practice in the stated context (i.e. the interpretation and prioritization of care values). The framework orients the ethicist and design team to the ethical issues demanding attention from the care ethics perspective.

The ethical questions and issues for different robots used in different practices by different users are going to be varied; however, this does not undermine that there are certain components that must be addressed in *every* instance in which a care robot is used. In every instance we must understand who the direct and indirect actors involved are and how they will be impacted. In every instance we must understand the care practice in terms of how values come into being (rather than understanding the practice in mechanical terms only). In every instance we must understand the context we are speaking of and the relationship this context has with the interpretation and expression of values. Accordingly, the framework is meant to draw the design team's attention to certain components and to show them how they are to deal with these components. The framework is not intended to say that each robot will undergo the same evaluation but rather, each robot will be evaluated according to the same criteria (i.e. the components of the framework).

Table 1.1 Care-centred framework for the ethical evaluation of care robots

Context – hospital (and ward) vs. nursing home vs. home setting …
Practice – lifting vs. bathing vs. feeding vs. delivery of food and/or sheets, collection of samples, playing games, etc. …
Actors involved – human (e.g. nurse, patient, cleaning staff, other personnel) and nonhuman (e.g. care room, mechanical bed, curtain, wheelchair, mechanical left, robot …)
Type of robot – assistive vs. enabling vs. replacement
Manifestation of moral elements – Attentiveness, responsibility, competence, responsiveness

Note: Table also found in van Wynsberghe 2013 and van Wynsberghe 2014.

Each of these elements have been judiciously chosen based on an analysis of the necessary and sufficient fundamentals for good care. This analysis is done from the care ethics perspective in Chapter 2. A detailed explanation of the elements and the justification for their place in the framework is provided in Chapter 5.

Why Design?

Discussing robots in terms of their "design" and the "design process" from which they result, demands an understanding of what I mean by both design and design process. By design I neither refer exclusively to the external appearance of the robot nor exclusively to the software programming of the robot; rather, to a combination of the appearance and capabilities of the robot. Of course the capabilities of the robot result from the programmed computer code and thus programming is subsumed within the element of capabilities. Appearance refers to the robot being humanoid, machine-like and/or creature-like as well as the morphology of the robot – the form and structure of the robot (e.g. does the robot have arms, does it have legs or wheels etc.).

In contrast, Feng and Feenburg describe "design" as a "process of consciously shaping an artefact to adapt to its goals and environments" (Feng and Feenberg, 2008, p. 105). This process of shaping the artefact is what I refer to here as the design process and not to design. My insistence to focus on design and the design process separately and as one being the process and the other the result of said process, rests predominantly on the relationship between artefacts and morality conceptualized in the ethics of technology and Science and Technology Studies (STS) domains.

Design and Morality

For some, artefacts are believed to have a kind of morality. Oosterlaken conceptualizes this morality in terms of a technology's ability to "expand capabilities" (Oosterlaken, 2009). This morality, or moral impact if you will, is a result both of the designers' intentional decisions as well as the technology's place within a network. I reference the term "network" intentionally to relate to Latour's approach known as Actor-Network Theory (ANT). For Latour, a network describes an amalgamation of human and non-human actors which interact together for: moral decision making, establishing norms and meanings and, determining outcomes. Actors are both human and non-human, thus a robot may also be considered an actor.

For some scholars in the field of STS, the morality of the artefact is accounted for through the phenomenon known as domestication; in short, the impact the technology has once it becomes an actor in a network of other human and non-human actors. Hence, domestication studies build on the concept of the network and the interactions between human and non-human actors (the material environment). This impact is observed and studied in terms of the meaning the technology takes on, how this meaning is established, how the technology propagates or alters existing norms, and lastly, in terms of how the technology prioritizes and interprets values. Given the technology's ability to maintain or shift an established morality, the artefact itself is said to be an actor. Actor-Network Theory insists on a lack of subjectivity or a homogenizing of the responsibility attributed to actors in a network whether they be human or non-human (technologies, the material environment, etc.). As such, according to ANT a care robot and a human care giver may have the same moral responsibility for the consequences of actions.

Structural ethics, on the other hand, is a concept in ethics of technology which maintains the concept of the network and the emphasis on the interactions between actors in a network, but adds other crucial dimensions for the analysis in this book. First, interactions occur among different networks on both the micro level as well as the macro level (the macro level refers to the overall institution or structure within which other networks exist). Second, responsibility remains in the exclusive domain of the human actors (Brey, 2014). Non-human or material actors are recognized as having a moral impact on the network, and for this reason, are referred to as moral factors. They factor into the moral decision making of humans; they are a factor in the establishment of traditional and/or new norms and values, and they are a factor in the establishment of the meaning attributed to a practice.

They are a factor because the artefact bears an impact on the decisions as well as the outcome of those decisions; however, they are not an actor because technologies are not capable of being "responsible" for their moral impact. Placing blame and/or praise on the "responsible" agent is a necessary condition for attributing responsibility to an actor. This is of no consequence to a robot and thus it is not possible to proclaim the robot responsible. Accordingly, the structural ethics approach concludes that a technology is still recognized as having an ethical impact but in light of it not being able to take responsibility the technology remains a moral factor and the full moral agents, capable of taking responsibility, are the (human) actors.

Thus, through design, a kind of morality is manifest. This morality comes about as a consequence of the designers' decisions, and is embedded into the robot. In Chapter 7 I will address the question of a robot's moral status and the relationship this has with its design (i.e. the selection of capabilities given to the robot). For now, let us agree that the care robot will invariably shape the decision making and actions of nurses, patients and other healthcare workers and thus establishes a new morality within the network or reinforces an existing one. This new morality comes not from the robot as a moral agent or actor –

capable of moral decision making – but from the robot as a moral factor. The degree to which the care robot will have an impact – decided by the role the robot is delegated – is an attribute decided upon throughout the design process and finds itself in the resulting design of the robot. So what can we say about design processes?

The Design Process

For Vincenti, a design process may be divided into either a normal or a radical one. A normal design process is one for which the "operational principle" and "normal configuration" are known and employed. The operational principle refers to how the device works (for example fluorescent vs. incandescent light bulbs have different operational principles). Alternatively, in radical design processes, "the operational principle and/or normal configuration are unknown or a decision has been made not to use the conventional operation principle and/or normal configuration" (Gorp and van de Poel, 2008, p. 79). An example of this would be battery operated cars in contrast with traditional cars.

Within a normal design process are regulative frameworks based on the operational principle and normal configuration. Such a framework describes "the system of norms and rules that apply to a class of technical products with a specific function" (Gorp and van de Poel, 2008, pp. 79–80). The framework "consists of all relevant regulations, national and international legislation, technical standards and rules for controlling and certifying products. It is socially sanctioned, for example by national or supra-national parliament such as the European Parliament, or by organizations that approve standards" (Gorp and van de Poel, 2008, p. 80). In a random design process no such framework exists.

For robots outside of the factory, no regulatory frameworks exist at present and thus designers resort to radical design processes. Such design processes are radical given the differences between robots manufactured in the factory and robots manufactured outside the factory. Firstly, the difference in performance environment – the factory is predictable and structured while the hospital or home is not (as) structured or predictable. Secondly, the difference in human contact – robots in the performance environment of the factory remain somewhat isolated while robots in the hospital will inevitably come into direct and indirect contact with humans on a daily basis. Thirdly, the size and capabilities of the robots – robots in the home or hospital are on average smaller than those used in the factory, with a wider range of capabilities and a more sophisticated skill level. Lastly, the materials used to create the robots – robots in the hospital will need to be sterile, for example.

Given that robots outside the factory will come into contact with humans much more often and in an unpredictable manner, the same safety standards used for industrial robots cannot apply. Accordingly, since industrial robots are used for different tasks than robots in the home, the same ethical considerations cannot apply for both.

Normal design processes follow socially and legally sanctioned ethical standards, and therefore the public is inclined to put their trust in designers and the resulting technical artefacts. Alternatively, in radical design processes, the foundational steps may not be in place, or followed – designers may not explicitly pay attention to ethical criteria. Conversely, with greater freedom in design, designers may pay greater attention to the ethical considerations at stake. This book holds the hope that designers of care robots may in fact pay greater attention to the ethical considerations, and their resolution, presented here.

Aside from the distinction between normal and radical design processes, there are hundreds of known processes, e.g. contextual design, user-centred design, use-centred design etc. (Dubberly, 2004). Essentially, a design process is a way of designing, of learning what the problem is, breaking it down into manageable fractions and deciding from this the best way to resolve the problem. Design processes typically involve a series of stages or phases during which the problem is deconstructed and the potential solution is proposed and worked into a prototype. Through each process, values are selected (both explicitly and implicitly) for embedding in the system.

When regulatory frameworks aren't available, design teams refer to internal design team norms, context, users, or the ergonomics of use. Although designers ought to observe and address the needs of stakeholders in context, this is not always what happens in practice (Dubberly, 2004, p. 28). Such a line of thinking reaffirms the work of STS scholar Madeline Akrich who claims that designs are the result of assumptions an engineer has of a context, rather than an understanding of the context in real life. It is for this reason that designers of late have embarked on understanding practices in context as a way of overcoming this discrepancy. In the case of VSD, Nathan et al. (2009 make the suggestion to understand values in context. Based on this, values are conceptually understood from a philosophical perspective but are also understood in terms of their manifestation in context.

Given the early stage of care robot development, explicitly addressing the design and design process of care robots is both essential and timely. In the specific case of care robots, the question is how the design process ought to proceed, given that it is a radical one. Without a regulatory framework to guide the design of care robots, the CCVSD approach presents a (radical) design process of sorts for an enhanced ethical focus.

The Ethicist as Designer

To accomplish the goal I have set out to do – incorporating ethics into the design process of a care robot – I am also making the claim that an ethicist must be included as a collaborator in the design process to help shape the future technology (van Wynsberghe and Robbins, 2014). This picture of the robot´s design process, with the ethicist as part of the design team, is still in its infancy and has yet to be fully articulated (van Wynsberghe, 2013a). It is my aim in this book to provide a concrete method for including not only ethics but the ethicist as well, into the design process of a robot in healthcare.

By showing the complexity of the ethical analysis required throughout the care robot's design I will provide reasons why an ethicist must be present rather than delegating the ethical analysis to an engineer or designer. Moreover, my aim is to outline the range of questions the ethicist must ask in order to accomplish his/her role on the design team.

The ethicist's role is to shape the design process and the resulting design of a care robot in a responsible manner. The impact of the ethicist is dependent on the context in which the care robot is being developed (academic vs industry), the stage of development (earlier on in the design process or later on), and the design team's goals, assumptions and values. The ethicist is evaluated both in terms of the successful inclusion of value analysis into a particular design as well as the inclusion of ethics into the practice of care robot design in more general terms.

Conclusion

The projected shortage of healthcare workers and resources has propagated a range of potential solutions to deal with the anticipated consequences. As previously stated, one of the proposed solutions is to introduce robots into care contexts (care robots) as a way of aiding care workers in institutions and/or as a way of helping care-receivers remain in the comfort of their own home for a longer period of time before entering into a care facility. Such a technology demands appraisal from an ethics perspective given its projected position in the moral practice of care.

From a care ethics perspective, the main question at stake has to do with the shift in roles and responsibilities following the introduction of the robot. Similarly, for those from the field of ethics and technology, the link between design and morality demands that design be the place to begin an assessment of the impact of said robots. Given the lack of socially and politically sanctioned design standards, ethical criteria in the design process of care robots may be jeopardized. Moreover, once a care robot has been created, no ethical guidelines for its retrospective evaluation exist to date.

In response to the lack of tools for the ethical evaluation of care robots both retrospectively and prospectively, I am creating an approach to do just that. The CCVSD approach is intended to be used in a prospective manner to guide the design of future care robots. Added to this, the CCVSD approach can also, as I will show in chapter five, be used for retrospective evaluations of current care robot designs. The focus on design does not presuppose a blind acceptance of all care robots for all tasks but is meant to reiterate the significance of design and the design process in the resulting artefact.

Each component of the framework is there for a very specific reason and highlights other elements which are integral to the care ethics perspective. In order to understand the significance of these components, we must take a closer look at care in context: what does it look like and what does it mean?

Chapter 2
Understanding Care in Context

Introduction

The task for which a robot will be used is one of the most important components for designing and programming the robot. It follows then that understanding and describing precisely the task for which the robot is to be designed is of great significance (Asaro, 2006; Engelberger, 1989). In the late 1960s, roboticist Joseph Engelberger proposed the idea of using robots in multiple domains outside of the factory (Engelberger, 1989). The way Engelberger's book proceeded, and the manner in which he suggested all robot applications ought to proceed, was concerned with outlining the capabilities of robots needed in order to fulfil the range of tasks to be delegated to the robot. If a roboticist envisioned a robot gas attendant then according to Engelberger the next phase was to outline what is required to fulfil this task: manual dexterity for handling the gas pump and planar locomotion to travel from the car to the pump, as examples. Next, the roboticist designs/programs the robot with the capabilities necessary to meet this mechanistic description of the task. In this way, roboticists could clearly envision the capabilities required for fulfilling certain tasks and accordingly what the design of the robot must entail in order to make this a reality. Engelberger's view presupposes that tasks can be understood as, or broken down into, a linear commodified process.

Other tasks, like "caring for the elderly", are not as straightforward as Engelberger may have thought. Care for the elderly could mean anything from having coffee one-on-one, helping an individual put on their shoes, bathing, lifting or feeding (eating assistance). What's more, the idea of understanding care tasks precisely, involves more than a description in mechanistic terms (the exact actions the robot will have to perform) as Engelberger suggested. Understanding the tasks for which a care robot is to be designed must also acknowledge the relationship such actions have within the holistic picture of care.

From the care ethics perspective, this holistic vision of care is crucial for evaluating a care institution and its ability to provide good care. Care should not be viewed as a commodity or as a series of unlinked actions done to meet standardized needs. These two threats described by care ethicist Joan Tronto that are among the seven warning signs of bad institutional care and ought to be avoided at all costs (Tronto, 2010). In order to provide good care, care activities should be seen as practices linked together. These practices bring together both caring actions with caring dispositions. Through care practices, as we will see in this chapter, the values that form the foundation of the healthcare tradition, are made real. The question then becomes whether or not it is possible for the design of a care robot to assist in the "making real" of said care values.

The following chapter begins by exploring care as a value on its own, with insights from the care ethics tradition. Most notably, I take the contributions of Joan Tronto in terms of conceptualizing care as a practice with phases and corresponding moral elements. Using a top-down approach I explore the realm of institutional care and the values of ethical importance within. This is done by examining and relating the abstract values of the WHO with the institutional values found in hospital policies and guidelines.

The chapter continues with an exploration of care practices to elaborate on the meaning of a care practice. This is done by presenting an example of a care task, outlining the range of actors (human and non-human) involved and how they interact, to show how this task is viewed as a practice rather than as a "task". Care activities are considered practices rather than tasks for multiple reasons; given their place in the institutional setting against a background of values, norms and assumptions; given their role in meeting the multi-layered needs of patients and nurses; given the range of abilities the nurse must embody; and, given the entanglement of meeting social and physical care needs through one practice. Understanding the meaning attributed to, and the complexity of, care practices may be the most significant finding for roboticists to understand the gravity of their design choices.

Unpacking the Concept of Care

Care may be one of the most difficult concepts to articulate. This is in part due to the ubiquity of the word but is also largely a consequence of the fact that one is assumed to know what care means given its revered place in many cultures. The work of Warren T. Reich nicely outlines the broad range of meanings and connotations care has embodied going back as early as Ancient Greece (Reich, 1995): care considered an essential ingredient for the preservation of humanity, "the key to the process of becoming truly human"; care consisting of helping acts directed toward healing; care referring to the bearing of responsibility or a burden, "worrisome or anxious care"; and care as a subjective experience, "the capacity to feel that something matters" (Reich, 1995).

Regardless, of how one perceives or defines care, care is still valued as something above and beyond simple care-giving tasks. It has a central role in the history of human kind. It is linked with concepts of good and theories of the good life. Recognizing the needs of another and acting on those needs is what we may call care in its most rudimentary sense, and it is this series of events that bestows a valuation on the care-receiver. Thus, care is a value in so far as it signifies the value of others.

The idea of caring as respect for human dignity is a common theme discussed in the literature. Care is described as "the humanity in one being reflected in the other" (Buber, 1958). Care has also been described as the recognition and response to the dignity of another; "the human act of caring is the recognition of the intrinsic value of each person and the response to that value" (Schoenhofer, 2001). As such, "care is not merely warm feelings or positive regard but is the outward expression or communication of those feelings" (Schoenhofer, 2001). As such, to care presupposes the recognition of the inherent worth of another.

Care is also understood as a verb. Caring may actually be divided into the idea of caring about and caring for. The dimension of caring about in the medical field implies a mental capacity or a subjective state of concern. On the other hand, caring for implies an activity for safeguarding the interests of the patient. In other words, it is a distinction between an attitude, feeling or state of mind vs. the exercise of a skill with or without a particular attitude or feeling toward the object upon which this skill is exercised. Caring for is unique in that it (often) requires the physical presence of the one exercising the skill for the benefit of the patient. Caring about does not place this requirement on the individuals involved in the scenario.

In 1982, the work of Carol Gilligan brought the significance of care in ethical decision making to the fore. As a result, care has been given special attention and value in the forum of ethical deliberation and decision making – nurses, bioethicists and care ethicists have become the stewards of care in the healthcare domain (Vanlaere, 2011; Pellegrino, 1985; Gilligan, 1982; Wilson, 2002). In fact, in many healthcare settings, reflecting on the process of care, or appealing to a notion of care, is the norm for ethical decision making. When introducing a new technology into healthcare, claims are often made about improving the level or quality of care. Accordingly, when evaluating the introduction of a new technology in healthcare, it is often done in terms of the impact on the provision of care (Wilson, 2002; van Wynsberghe 2008; 2009; 2012; 2013). From this recognition of care in healthcare and its lack of acknowledgement in traditional ethical theory, the care ethics tradition evolved.

Care ethics is thought of, not as a pre-packaged ethical theory ready to use in a given context, but as a perspective or stance from which one can theorize ethically; a lens from which one may begin moral deliberation in the care of others (Verkerk, 2001; Little, 1998). This lens from which one begins to theorize is coupled with the direction in which such theorizing ought to take place. For many care ethicists, the care ethics stance demands a recognition of the relational status of persons along with a focus on responsibilities rather than rights (Tronto, 1993; Tronto, 2010; Little, 1998; Noddings, 1984; Buber, 1958). This focus on responsibilities over rights does not exclude a discussion of care values; rather, care values are understood as being responsibilities of the care institution as well as the care provider.

The Values in Care

Alternative to the idea that care in itself is a value, linked with the good life and with a valuation of another, is the idea that beneath the umbrella concept of care exist many other care values. These values are given importance for their role in care; in making care what it is. It is through the manifestation of these values that one comes to understand what care really is in practice. It is therefore fruitful for the topic of embedding values to understand these values and their link with outcomes in a healthcare context.

I begin with a top-down approach in which I look to the values articulated by the governing body of healthcare, namely the World Health Organization. The WHO framework for people-centred health narrows in on the values in healthcare stemming from the patient's perspective; *patient safety*, *patient satisfaction*, *responsiveness to care*, *human dignity*, *physical well-being* and *psychological well-being*. This is not to say that other values like innovation or physician autonomy are not valued but rather from the patient's perspective, the listed values are the ones with the greatest ethical importance and will thus be used in my evaluation of implementing robots in the care of persons.

The above values are meant to structure and guide the overall tradition of healthcare. But such values must also be specified when understood in a more specified context – at the institutional level. In support of the values identified through the WHO guidelines and mission statements are created for hospitals and nursing homes across the globe. In London Ontario (where I was fortunate enough to spend time observing and volunteering) the guideline of hospitals includes additional values like: *compassion*, *integrity*, *dedication*, *respect* and *accountability*. These values relate to the manner in which care is provided (i.e. with integrity and accountability) rather than the ends to which care aims (i.e. the WHO value of physical wellbeing or patient satisfaction). They highlight the importance not only of caring actions but also of caring dispositions. In short, the guidelines for a hospital ward or a nursing home aim to specify the more abstract values presented by the WHO but they also aim to account for any missing elements.

To further specify the values expressed in hospital policies and guidelines, the College of Nurses (in Ontario) (CNO) has created an ethical framework.

This framework "describes the ethical values that are most important to the nursing profession (in Ontario)" (Nurses of Ontario, 1999, p. 3) (see Table 2.1). Thus, as we continue to work our way down from the WHO's list of values to the values of the institution, there are still missing values that define the roles of carers in their profession. Interestingly, it is only when the values of the nurse are listed that the value of privacy is stated and we are all aware of the significant role that privacy plays in care. One would assume that such a value is not stated by the WHO but is subsumed under another value like patient satisfaction or human dignity.

The reason for listing all these values is not meant to exhaust the reader but rather is intended to point out the difficult goal of the ethicist who wants to suggest incorporating values into the design of a care robot. Which values does one choose? Whose interpretation does one rely on? Furthermore, how does one suggest that an abstract value such as human dignity should be embedded into the design of a care robot? To help answer this I turn to the care ethics concept of a care practice.

Table 2.1 Values according to the ethical framework for nurses (in Ontario)

Value	Definition
Well-Being	Facilitating someone's good or welfare and preventing or removing harm.
Client Choice	Client choice means self-determination and includes the right to the information necessary to make choices and to consent or to refuse care.
Privacy	Limited access to a person, the person's body, conversations, bodily functions, or objects immediately associated with the person.
Confidentiality	Involves keeping personal information private. All information relating to the physical, psychological, and social health of clients is confidential; as is any information collected through nursing services. Clients have the right to confidentiality, and nurses make an implicit promise to maintain confidentiality.
Respect for Life	Human life is precious and needs to be respected, protected, and treated with consideration. Respect for life also includes considerations of the quality of life.
Truthfulness	Speaking or acting without intending to deceive. Truthfulness also refers to providing enough information to ensure the client is informed. Omissions are as untruthful as false information.
Fairness	Allocating healthcare resources on the basis of objective health related factors.

Care Tasks versus Care Practices

A Care Task

A predominant threat in institutional care is the conception of care as an isolated action or task without recognition of its place within the process of care or its relationship to other actions/tasks. "Care institutions have to think about the nature of the caring process as a whole in order to guide their actions" (Tronto, 2010, p. 162). Thus, the concept of a care task for which the robot is to be designed must be abandoned and in its place I introduce the concept of a care practice. To show the difference between a care task and a care practice I take the example of bathing, carried out in the hospital or nursing home. I have specifically chosen to look at this activity given its recognition as a moment in care that much more significant than the mere bathing of the corporeal dimension.

Although I describe the bathing of a patient in their room it is true that oftentimes patients with more mobility will be taken to a tub room for bathing. I use this practice and the description thereof to set the stage for further analyses of practices using the Care-Centred framework in future chapters. What's more, it is also a practice for which care robots are currently being designed (Sanyo bathtub), adding weight to the understanding of the practice of bathing in its totality.

> ***Bathing a patient in their hospital room***
> The human actors are the nurse and the patient. The non-human actors are the mechanical bed, the sink in the room, the cleaning supplies, the curtain to enclose the patient, the window in the corner of the room, and the door (the list is exhaustive but we understand the range of non-human actors). The practice begins with the nurse entering the patient's room, indicating what she will be doing (cleaning the patient), asking if that is ok and beginning to organize the cleaning supplies in close proximity to the patient. When things are set-up, the nurse encloses the curtain around the patient and begins to undress the patient. Using a wash cloth, the nurse cleans all orifices of the patient's body asking if the temperature of the water is fine and other personal items like how the patient is feeling. The nurse is also checking the patient's skin colour and temperature as indicators of well-being.

The task of bathing, as with many other tasks like lifting or feeding, can be described as goal-directed in the sense that its aim is toward the exclusive goal of bathing or lifting. To achieve this goal there are mechanistic functions that need to be satisfied, namely, the patient must be lifted at a certain speed, angle and force and placed in the wheelchair. This is the manner in which Engelberger would have described the task. Describing the activity in this way makes it easy to envision a robot integrated into the task. But the task of bathing, as with many other tasks, can also be described in its full sense to account for the development of caring skills that occurs throughout the activity, i.e. learning how to test the patient's physical and neurological status. Additionally, it can also be described in a way that recognizes this activity as a moment in which the therapeutic relationship is formed and/or strengthened. These added dimensions make it problematic to conceptualize such an activity as merely a task. Instead, I use the concept of a care practice to show the richness and interconnectedness of all the activities that occur in the healthcare domain.

The Meaning of a Care Practice

Within the field of care ethics, the idea of conceptualizing care in terms of a practice is what allows for care to be considered as the marriage between action and disposition. This fact alone adds another dimension to the meaning of a care practice; a care practice is about much more than the mechanical task or care activity. The challenge then is to capture the meaning of a care practice so it can be used for a robot's design. To do this I will describe the many variables, elements and interactions involved.

> The notion of a care practice is complex; it is an alternative to conceiving of care as a principle or as an emotion. To call care a practice implies that it involves both thought and action, that thought and action are interrelated, and that they are directed toward some end. The activity, and its end, set the boundaries as to what appears reasonable within the framework of the practice. (Tronto, 1993, p. 108)

In a similar manner, I refer to care practices to indicate the complexity of actors, needs and values involved. I define a care practice as the attitudes, actions and interactions between actors (human and non-human) in a care context that work together in a way that manifests care values: a care practice facilitates the realization of care values. All of these factors contribute to the meaning that a care practice takes on.

What is also clear from Tronto's definition is that the end is not the only means for evaluating a care practice; both thought and action are interrelated. Thus, a care practice is defined by the interactions between actors but also *how* these interactions take place; care practices are values working together. Take the WHO value of patient safety as an example; the value of safety is realized through the (protective) actions on the part of the care professional. This could be wearing sanitary gloves, washing ones hands, double checking the dosage of medication assigned to a patient and so on. How one does these activities realizes different values: making sure that one does such activities on a regular basis or thoroughly (washes their hands) is the realization of the values of accountability and dedication.

There are human actors, the nurse and patient predominantly, but at times there may be a host of other staff contributing to the practice. There are also non-human actors in the network; the hospital or nursing home room, the TV, the mechanical bed, the mechanical lift, the telephone, the respirator, the window, the door, and so on and so forth. The care values are realized through the interactions of all the elements in the practice, the humans and the material environment.

For instance, in the bathing scenario presented above, all actors (human and non-human) interacted and were important for decision making throughout the practice. The nurse's decision to enclose the curtain was impacted by the fact that the curtain was there and by the fact that the door would remain open to the hospital ward and so on. Added to this, one may presume that the curtain enclosed around the patient is a manifestation of the value of privacy. But, the door to the patient's room was open and the window was not closed. Thus, partial privacy is achieved, but not full. However, it is possible to suggest that complete privacy may be seen as putting the patients at risk of abuse. The privacy required for this practice is from other patients and casual passers-by.

For nurses, the practice of bathing is highlighted as a moment in which the nurse gains additional information as to the patient's medical status (Pols, 2004). The nurse assesses the patient's physiological status, neurological status, temperature of the skin, and overall sense of the patient's mood and/or recovery. The nurse learns of the patient's preferences for types of soaps or temperature of water. What's more, it is an opportunity for the nurse and patient to engage in social conversation and essentially to build the relationship between the two thereby establishing and/or maintaining trust through the nurse's commitment to privacy, confidentiality and compassion.

Perhaps the most significant part of this practice is the role it plays in establishing and/or maintaining the therapeutic relationship. Building a trusting relationship is important not only for the next time the nurse needs to bathe the patient but also for additional practices the nurse will be engaged in (Nurses of Ontario, 1999). Later on in the care of the same patient the nurse requires that the patient be honest about their symptoms, comply with their care plan and take their medications. Without a rapport or trust between nurse and patient, the patient is not as likely to meet these needs of the nurse.

Values Beyond the WHO

In addition to the values already discussed there are other values of equal importance that have not yet been listed. These are values that become apparent when the care practice is described as it occurs in context: values like eye contact when the nurse is talking to the patient; values like human presence when the nurse enters the room for the duration of the practice. These are both values not listed in any of the guidelines presented here but are nonetheless essential in the provision of good care. Eye contact plays a role in a patient's assessment of the truthfulness of the nurse. Human presence plays a role in conveying the worth and dignity of the care-receiver.

There is also the value of "tinkering" coined by Annemarie Mol (Mol, 2010). Such a value refers to the nurse's ability to alter his/her behaviours and actions based on the changing needs of the patient and the nurse's perception of such changes. This demands a sense of attentiveness on the part of the nurse and presupposes the nurse is physically present to perceive such changes. The nurse's responsiveness will then take into consideration any change in the patient's physiological status such as temperature of the skin, etc.

All of these aspects bestow meaning to the practice and illustrate how the practice of bathing is much more than the corporeal dimension of cleaning a patient's body. Jeanette Pols speaks to this in great detail and discusses the bathing of patients on psychiatric wards and how the practice reflects a conception of the patient as a citizen (Pols, 2004). Pols identifies four different washing repertoires, the first repertoire is labelled, "washing as part of individual privacy". In this repertoire patients choose how often and when washing occurs and thus individual preferences and tastes are key; "by relating to a patient's history of washing, the activity can be tailored to what he or she is used to and prefers" (Pols, 2004, p. 50). In the second, "washing as a basic skill", the bathing of an individual is used as a moment in which they are empowered to keep their own basic motor skills in tact (Pols, 2004, p. 55). The role of the care-giver in this instance is to encourage and motivate the patient to fulfil certain portions of bathing on their own and to assist in the parts of washing that the patient cannot do themselves.

The third repertoire, "washing as a precondition", outlines the practice of bathing as a step towards the self-actualization of the patient (Pols, 2004, p. 60). The practice of bathing is used to develop the patient's autonomy; "autonomy in care for self-actualization is a matter of developing and choosing rather than doing things oneself (basic skills) or being free to decide on one's individual life (privacy)" (Pols, 2004, p. 61). Repertoire four, "washing as a relational activity", illustrates the practice of bathing as an intentional moment for establishing relations as the goal of the care. Thus, learning basic skills is not the primary aim but rather being connected with others is. As such, the nurse becomes a part of the social network of the patient.

In describing this repertoire, using an example from her own fieldwork experience, Pols addresses the aspect of communication: "communication is not always easiest by 'talking'. On the contrary, (he perceives) verbal communication is often difficult. Washing creates an alternative situation for communication. There is a clear task at hand and 'the rest comes with it'" (Pols, 2004, p. 67).

Thus, the end goal of bathing differs depending on the patient and their particularistic condition and history. As seen through Pols' work, when the end goal of the practice changes (developing basic skills vs. developing a relationship between care-receiver and care-provider), the prioritization of values changes along with it. Thus, not only is a practice like bathing recognized for its utility in meeting the corporeal needs of a patient but it is also granted a deeper meaning when we observe the expression of values and the meaning the practice takes on as a result of such expressions. This understanding must be made explicit in order to begin to understand how a care robot might impact the provision of care when introduced into a practice like bathing.

Naturally the context in which the practice takes place plays a role in the transition from task to practice. As was just shown, all of the actions of the nurse are aimed at fulfilling the socially and legally sanctioned values of the tradition of healthcare found in the guidelines of the institution and the professional codes of conduct. Added to this, the ends that the practice serves, not only in terms of manifesting values but in the meeting of needs, adds a further complexity to the concept of a care practice and it is the element of "needs" that I will explore next.

The Multi-Layered Needs of Patients

Care is thought of first and foremost as a response to the needs of another. For Tronto, this marks the starting point of care. Additionally, needs act as the thread linking all actors and elements of the care practice. As humans we are constantly in a state of need, the differences occur in the amount of need one individual requires and the manner in which these needs may be met; some individuals require more assistance than others for their whole life while others require additional assistance for only a short period of time. Thus, to begin, we must first understand needs in a cyclical and fluctuating way.

It follows then that the response to needs is also a process vs. a single act in time. Responding to needs in the process of care articulated by Tronto refers to the following: attentiveness to the needs of the other, taking responsibility for the needs of the other, embarking on an action to meet the needs of the other and engaging in reciprocity with the patient to assure their needs have been satisfied. Maintaining the needs of the other as the central focus of the care process reflects the concept of "ethical sensitivity", a skill of nurses essential for the provision of high quality care (Weaver, 2008).

This last point, directs our attention to the idea that not only do nurses, and other healthcare workers, have needs in terms of resources but there are needs of the nurse fulfilling their role as nurse; skills or attributes of the nurse that are needed in order to provide good care. The remainder of this chapter is meant to outline the concept of needs within the healthcare tradition. This is not an easy task, however, determining needs is described as "one of the foremost political struggles of any account of care" (Fraser, 1989). In fact, "any agency or institution that presumes that needs are fixed is likely to be mistaken and to inflict harm in trying to meet such needs" (Tronto, 2010, p. 164). Consequently, needs must first and foremost be recognized in all their complexity. Added to this is the idea that the complexity of needs comes not only from the way in which they ought to be met but rather from the recognition that needs are dependent on a unique individual at a given point in time.

Conceptualizing Needs

Although need is often spoken of in the economist sense, in terms of want, need ought to be distinguished from want. Want brings with it a connotation of voluntary conduct; I desire something even though I can live without it. Need, however, brings with it a connotation of conditions or things that are necessary but lacking, or conditions requiring relief. Linked with this, is the idea of conditions or things being necessary or vital.

What we think of as "needs" changes. They change over time for particular individuals, they change as techniques of medical intervention change, they change as societies expand their sense of what should be cared for, and they change as groups make new, expanded or diminished demands on the political order. The demands placed upon institutions change (Tronto, 2010).

Thus, needs in the healthcare tradition must be conceptualized in terms of the meeting of needs of individual patients, the needs of care-givers as well as the needs of the larger institution. For the latter, the institution needs support staff, managerial staff, equipment, insurance and so on and so forth. For the former, needs of an individual patient must be worked out according to that individual person, their medical history and the status of their condition.

Needs of the care-giver cannot go unnoticed for without these care could not be provided, or at least could not be provided well. The care provider has needs for the institution to meet in terms of resources; needs in terms of the patient they are working with; and, needs in terms of what is needed of them as care professionals in order to provide good care (i.e. skill, competence, compassion, empathy). Each care provider cares for different patients, fulfils different roles and responsibilities and is themselves a unique individual.

What all the concepts have in common is the multi-dimensional character of needs based on the multi-dimensional nature of the person, their relationships, and the institution they're in. This is in contrast to a vision in which the patient or care provider is objectified patient or the care is standardized. Knowing that everyone has needs and that these needs will change depending on the context and situation is not a new phenomenon, but how are needs actually met? Meeting needs isn't quite as clear.

Meeting Needs

When needs are thought to be met through care actions alone they are conceptualized as a commodity. In such a picture needs are met through the fulfilment of tasks without a sense of the overall process of care or a concern for the disposition of the care-giver. Care-receivers are conceived of as clients and the market model is instilled. But such a commodification of needs often leads to alienation; who will have access to the care for their needs to be met? (Tronto, 2010). Oftentimes marginalized demographics will not. To overcome such health equity issues Amartya Sen and Martha Nussbaum propose the capabilities approach (Verkerk, 2001a; Nussbaum, 2000; Sen, 1985). According to this perspective needs are conceived of in terms of capabilities; the capability for play, the capability for relations; the capability for a healthy life as examples. Needs are not defined in terms of majorities but take into consideration those marginalized demographics without the same baseline capabilities. When such capabilities are met, said person is considered to have a functioning.

For Tronto, needs are met through a four stage process of care; "caring about (recognizing a need for care), caring for (taking responsibility to meet that need), care-giving (the actual physical work of providing care) and care-receiving (the evaluation of how well the care provided had met the caring need)" (Tronto, 2010, p. 160; Tronto, 1993). These phases are not in a linear order but are rather intertwined and happen concurrently in many instances.

Although one might claim that needs are tangibly met when the care-giver engages in an activity to satisfy the needs of the care-receiver (care-giving), meeting needs cannot be reduced to this phase alone. Meeting needs presupposes that needs have been identified correctly, that someone is taking responsibility for meeting these needs and that someone will assess whether the action taken has met the identified needs. It follows then that many actors (human and non-human) are involved in the meeting of needs, each delegated a certain amount and type of responsibility. In accordance, alongside the four phases of a care practice, Tronto identifies four corresponding moral elements.

The moral elements act as the buttress of care, the manner in which the ethical nature of care may be evaluated. I claim that through Tronto's account of the moral elements she is ultimately presenting a normative criterion for the evaluation of care practices.

Good care requires that the four phases of the care process fit together into a whole. Similarly, to act properly in accordance with an ethic of care requires that the four moral elements of care (attentiveness, responsibility, competence and responsiveness) be integrated into an appropriate whole. Care as a practice involves more than simply good intentions. It requires a deep and thoughtful knowledge of the situation and of all of the actors' situations, needs and competencies (Tronto, 1993, p. 136).

The element of attentiveness refers to the care-giver's ability to perceive the often changing, unique needs of the patient/person. The care provider in this role need not be the one dispensing care, their role may be only to address the initial needs of a patient to determine which doctor or nurse should attend to them. While their role ends here, their responsibility may not. The nurse in charge of admitting the patient will not follow the progress of the patient but may be responsible to ensure that the patient is directed to the next specialist. This ties in with the element of responsibility.

Responsibility as the second moral element means that an individual, or institution, be responsible for the needs of said patient. In terms of the institution, one might assume that when entering the institution of the hospital, nursing home, hospice, etc., the institution both symbolically and casually takes responsibility for the overall care of the patient. In the above scenario, the nurse who passed-on the patient to the specialist, we might suggest that she/he delegated the responsibility of physiological care to the specialist. The delegation of responsibility does not necessarily fit this linear model described; however, it provides a way of visualizing the chain of responsibility among healthcare professionals, the complex sharing of responsibility and the relationship roles and responsibilities share from one phase/element to another.

The element of competence is translated as skill. The care provider is required to fulfil their role and responsibility of care-taking in a skilful manner. If not, the responsibility assigned to the role requires that they be blamed, held accountable and/or liable and thus punished accordingly. Skill refers not only to the actions that the care giver carries out but also to the manner in which these actions are fulfilled. Thus, a care giver may skilfully bathe a patient without any verbal communication if he/she knows that this patient would rather not speak or be spoken to.

Reciprocity refers to the component in which the care-receiver is actively engaged in responding to their care provider and the care provided. This is not a one-time event but rather is continual throughout the multiple phases that occur. A patient may be responsive and engage in a reciprocal relationship any number of times during a patient's time in a care institution. What's more, the phase of responsiveness or engaging in a reciprocal interaction may also be conceived of as a need of the care provider – without a response, positive or negative, from the care-receiver, the care-giver will not know when needs have been met or whether they have been met to the satisfaction of the care-receiver.

It is important here to digress for a moment and discuss the issue of meeting needs from the perspective of ethics of technology. In so doing, it is important to acknowledge the relationship between needs, care and technology. Such a relationship is neither causal nor linear. One does not begin with needs isolated from technologies and then apply care with or without the use of technologies. Rather, needs are often the result of interactions between humans and existing technologies (Verbeek, 2011). For example, the need for surgical intervention is only recognized as a need when we have the technology to provide such an intervention. Without such technologies, the need would be considered a want. It follows then that the introduction of endoscopic tools has introduced a need for "minimally invasive surgery" whereas prior to the technology, conventional surgery was the only option and so conventional surgery was where the need ended.

It is also important to indicate that I do not believe all dimensions of needs *ought* to be met by human intervention alone. I see that as being neither possible nor desirable in our current healthcare situation. It is not possible because the meeting of needs happens in the context of the hospital which employs a wide spectrum of technologies for its functioning, from the mechanical bed, to the foetal heart monitor, to the TV. It is undesirable given that technologies should also be recognized for their benefits to care. For example, sterilization technologies allow for rooms to be kept uncontaminated. This is necessary in the treatment of individuals suffering from cancer and undergoing chemotherapy treatment. This provides a significant benefit to the patient and nurse and is not something that could be done without technology (granted the need for it also arises from another technology, chemotherapy and radiation treatment for cancer).

Thus, needs are met through a variety of actors (human and non-human), actions, attitudes, and roles that together express the values in care. Needs are met through care practices which may be broken down into the four phases of care each assigned a corresponding moral element used to evaluate its moral quality. Each of the moral elements must be attended to in striving for good care. The quality of a care practice will suffer if time is not allotted to the development of attentiveness of the nurse towards a patient or if the necessary skills are not cultivated to ensure the competent completion of an action. Through such an analysis of care, the complexity of identifying purposes, roles and responsibilities is revealed. It is this complexity and intricacies of roles, responsibilities, manifestation of values and interactions between human actors and the material environment that must be made explicit and criticized prior to the design and introduction of a care robot for a given practice. Consequently, this understanding of needs orients the recognition of needs and the multi-dimensional nature of said needs, as a high ranking value. Without this starting point, care cannot proceed or might proceed in a dodgy or hazardous way.

Paying tribute to the idea that needs are multi-layered, and thus care practices serve multiple ends, we end up in a discussion of the variety of skills and attributes the nurse ought to embody. The nurse is responsible for understanding needs in this dynamic, holistic sense but is also required to understand a range of communication forms (verbal and non-verbal) and to be able to "tinker" a care practice accordingly. What are also needs in a sense then are the abilities of the nurse (or other care-giver) to bestow care in a compassionate and empathetic manner. This refers to the capabilities of the nurse, as a need.

Abilities of the Nurse: Cultivating Care Skills

With an idea of what "needs" refers to, we cannot ignore that for good care, another need has to do with the capabilities of the care-giver. Care ethics and bioethics highlight the role of virtues and virtue ethics in the evaluation of care. The "good care-giver" is one that does so in a way that promotes their own moral development while at the same time fulfils the best interests of the care-receiver (Vanlaere, 2011). Each of the moral elements listed above point to a core component of the moral significance of the care-giver – care-givers are (traditionally) moral agents who assume moral responsibility for the care of an individual.

For Tronto, all of the moral elements essentially refer to capabilities or necessary skills of the care provider. Specifically, "the second dimension of care, taking care of, makes responsibility into a central moral category" (Tronto, 1993, p. 121). As such, moral responsibility is an important attribute for the care-giver to possess. This is because of the types of decisions the care-giver will have to make (those that carry with them moral consequences) but also because of the manner in which care ought to be provided (a caring disposition and a moral agent is required to comprehend the significance of this). Moral responsibility also assumes that the care-giver ought to exercise impartiality and justice when necessary and not rely on subjective emotions to guide their actions and decisions (Vanlaere, 2011).

Being in-tune with the delicacy of the situation, and how to address it (what Tronto refers to as attentiveness), can also be referred to as "ethical sensitivity" (Weaver, 2008) or "tinkering" (Mol, 2010). The former adheres to the idea of care as caring about while the latter is closely linked with care as in caring for, albeit they are not mutually exclusive.

There are other human attributes of the care-giver which are also necessary for promoting many of the values in care, namely the ability to empathize and portray compassion for the care-receiver. These are important emotions for the care-giver to portray; however, the care-giver must always be in a state of balancing the portrayal of emotions with an objective stance pertaining to the good care of the patient. The care-giver cannot allow emotions to interfere with what is best for the patient medically speaking. The care-giver must take a distance from the patient in this respect, while at the same time allowing the patient to understand that they take this stance from a disposition of care – of beneficence for the patient's best interests.

It must be stated here that for some, the use of a care robot, or any another technology for that matter, poses a threat along the lines of what it does to the care-giver in their role as carer as well as what the impact may be on the care recipient. "Critics fear, perhaps justifiably that care-givers might be less attuned to the specific needs of care recipients because a technological crutch is available" (Borenstein, 2011, p. 260). For Shannon Vallor, this aspect presents the main concern when discussing the prospect of robots in care (Vallor, 2011). Vallor claims that the use of care robots prohibits the cultivation of the necessary care skills mentioned above. What's more, that removing the care-giver from their role prevents the cultivation of certain human capabilities like empathy and compassion, necessary for the flourishing of the care-giver as a person engaged in relationships of their own.

I cannot deny this claim as I believe it to be true when and if care robots are designed without an understanding of the skills that develop through care practices. That being said, I also believe it is possible to sketch care practices in the detailed manner, the same that I have attempted to do in this chapter, and to outline very clearly where a robot could play a role without interfering with the development of skills. This is risky business because such an endeavour could backfire if care robots become a pervasive presence and indeed become the crutch the Borenstein mentioned. But isn't this the same with any other technology introduced into healthcare? How would the robot be so different?

The robot could be different if it were designed in a methodical way that demanded fine grained descriptions of care practices to ensure that the care values expressed and the skills developed were not interfered with but rather were enhanced. Not to worry, this is the aim of the Care-Centred Value-Sensitive Design Approach!

Conclusion

The goal of this chapter was to outline the many fundamental values within the care ethics tradition and the importance of understanding their relationship to one another, as well as how they are expressed in a care institution. The noteworthy contributions of care ethicist Joan Tronto become indispensable for understanding care as a practice with corresponding moral elements. Only by understanding values at the level of the care context and care practice, can we begin to speculate on the program requirements of a care robot. An additional goal of this chapter was to sketch the significance of the care practice, both in terms of understanding the meaning of the care practice as the forum in which values are made real, as well as understanding the relationship one practice shares with another and with the overall process of care. For these reasons I insist on referring to care practices as such rather than as mere tasks.

Added to this is the recognition that care practices serve more than one purpose or need. The simple act of asking "how one is doing today" means a great deal to the patient on a personal level; it reflects a valuation of the patient through their eyes. The act of bathing a patient with gentleness and compassion while asking personal questions and staying true to the preferences of the patient reflects again the valuation of the patient's dignity, integrity and respect for life. The act of telling jokes while one is serving a meal to elderly patients in a nursing home adds a light-hearted social dimension to a resident's day adding to patient satisfaction. Asking residents of the nursing home "what they would prefer to do today" grants them choice and preference in their care – a manifestation of their worth once again.

These practices add meaning to the care practice above and beyond the expression of institutional values, they emphasize a valuation of the client as a person rather than as merely a patient. In this way, we observe the link that care shares with the provision of well-being, of a good life: the values in care mirror the values which serve as the foundation for theories of the good life. From a care perspective, the good life is one in which persons are engaged in relationships and feel connections with others. In other words, care practices make real/tangible the values in care and thus take on a deep meaning.

One might wonder then about the meaning of the care robot once integrated into a care practice and suggest that its meaning arises within a care practice from the interactions with nurses and patients, in the value-laden context of a hospital or nursing home. The questions we are left asking now are multiple: will care practices still be considered such if the robot providing care does not possess the same attributes as a human care-giver, does not fulfil social care in tandem with physical care, does not address the multi-dimensional conception of needs, and does not project empathy and compassion throughout the care practice? Alternatively, we may find that a care robot has the potential to re-introduce values into a care practice that were lost at some point.

Before we can get to these questions, let us first take a look at the technology we've been referring to; the robots.

Chapter 3
Robots and Robot Capabilities

Introduction

With an understanding of the complexity of care values and care practices the goal now is to understand the technology in question. Just as care is an incredibly difficult concept to define, robots may be one of the most difficult technological innovations to define. This is partly because of the great technical knowledge to understand their functioning but also because of the role media has played in shaping the image of a robot in the minds of society. The images given by the media, represented by – Star Wars' C-3PO, Star Trek's Data, Pixar's WALL-E – all represent a class of robots not yet realized by today's technology. These futuristic human-like robots may be part of the future or may never be realized.

In the *Springer Handbook of Robotics* (Siciliano, 2008), it is acknowledged that there is no consensus on what a robot is today. Different definitions are given depending on the class of robots discussed or their application. Because of the wide variety of robots currently available, some authors discuss particular prototypes while others leave out a definition of robots altogether and instead focus on robotics – the study of robot foundations and methods, or the science and technology of robots.

In this chapter I explore what robots are, the difficulty in their definition, the variety of types and the range of capabilities a robot may possess. The goal of this chapter is to introduce the reader to the current state-of-the-art in robotics research. To do this, and to align with the overall aim of the book, I will describe the critical capabilities that need to be considered when designing care robots (e.g. safety, vision, mobility, grasping, force feedback and others). By presenting current robot capabilities and prototypes, the aim is two-fold: 1) to prepare the reader for evaluations of real world robots rather than speculative robots (Smits, 1995a; Rip and Nordmann, 2009); and 2) to educate the reader on the technical capacities of a care robot as a pre-requisite for plausible evaluations (Swierstra, 2007).

Defining a Robot

The word robot was coined by Karel Capek (1890–1938) in his play *R.U.R.* (Rossum's Universal Robots) where he used it to refer to a race of manufactured humanoid slaves (Capek, 1923). Robots, as they were first thought of, were considered machines that can do the work of humans. The term robot essentially replaced the terms android and automaton which had been traditionally used until that time.

Although the word robot was introduced in the last 100 years, society's fascination with robots, or artificial man-made creations resembling the biological, can be traced back to Ancient times including the clay Golems of Jewish legend or Talos, the man of bronze guarding the Cretian island of Europa from pirates in Ancient Greek mythology. In Renaissance Italy, Leonardo da Vinci sketched plans for a humanoid robot (Lin, 2011). In seventeenth- and nineteenth-century Japan, automatons like mechanized puppets were made. In eighteenth-century France, Jacques de Vaucanson made several life-sized automatons: a pipe player, a flute player and a duck. The mechanical duck could flap its wings, crane its neck, swallow and give the illusion of excreting food (the excretions were actually stored in a hidden compartment). Thus, although robots are thought to be a modern fascination history reveals this to be far from the truth.

When defining a robot, the "sense-think-act paradigm" (Borenstein, 2011, p. 259) is as close to consensus as one might find in terms of a robot differentiated from a computer system. Alternatively, some believe that robots are similar to computers up until the point of physical agency – "a robot physically embodies the link between perception and action" (Franklin, 1997). Others believe the distinction between robots and other appliances or devices is the element of autonomy – that robots are capable of completing tasks without direct human input (Thrun, 2004). Others believe the definition of a robot changes as the technology for creating robots develops.

The term robot differs from robotics; "the study and use of robots". (Mitcham, 2005, p. 1654), or "the science and technology of robots" (Siciliano, 2008, p. 1). The term robotics has also undergone changes depending on the state of the art. In the 1980s robotics was defined as "the study of intelligent connection between perception and action". This more sophisticated description of robotics indicates the underlying core definition of robots as being capable of both perceiving their environment and executing some type of action.

Despite the uncertainty of defining robots there are still certain attributes or characteristics shared among robots: all robots are man-made or artificial and are intended for the purpose of fulfilling a task for a human. All robots can sense their environment and can manipulate and interact with things in this environment. A robot is programmable, and re-programmable in many instances. A robot must have a high level of intelligence which affords them the ability to make choices based on the environment or a set of pre-programmed sequences of action.

This high level of intelligence comes in different forms and in varying degrees. Some robots are intelligent in the sense that they can manoeuvre through busy corridors while avoiding objects and people. Other robots are intelligent in that they can read and perceive human gestures and respond accordingly.

One class, or type, of robot that is often confusing is that of humanoid robots. For some, humanoid robots refers to those robots which "selectively emulate aspects of human form and behaviour and may come in a variety of forms from complete human-sized legged robots to isolated robotic heads with human-like sensing and expression" (Kemp, 2008). The latter description – with human-like sensing and expression – is often referred to as social or sociable robots. These robots are designed "to engage people in an interpersonal manner, often as partners, in order to achieve social or emotional goals" (Breazeal, 2008, p. 1349).

A common misunderstanding with humanoid robots is that they also possess the capabilities referred to for social robots (Kemp, 2008; Verrugio, 2008; Ng, Thow and Hing, 2009). The assumption is made that a human-like appearance presumes a set of social capabilities the robot must have. Kiesler and Goetz (2002) found in experiments that "the presence and absence of humanoid features and the behaviour of the robot influences people's assumptions about its capabilities and social inclinations".

In contrast to the more mechanical looking humanoid robots are androids; "android robots are designed to have a very human-like appearance with skin, teeth, hair and clothes" (Breazeal, 2008, p. 1351). These robots are most commonly used to test Masahiro Mori's theory of the uncanny valley (Mori, 1970) – when robots resemble humans to a close degree it elicits feelings of revulsion and discomfort among observers. "Once the robot reaches a point at which their resemblance is close to perfect but eerily dissimilar enough such that we no longer trust them – that sudden shift in our affinity is represented by a dip or valley on the curve. But the trust returns as the anthropomorphism approaches perfect resemblance to human appearances" (Bekey, 2011, p. 25).

Robot Capabilities and Features

The following section is intended to draw the reader's attention to the variety of capabilities a robot may have. The sub-section titled, "Capabilities for Safe Interaction with Humans", discusses the feature of robot safety in terms of interacting with humans in a human environment, hence a robot outside the factory. Programming for these conditions may be done in a variety of ways as discussed in the section. This section will also touch on the standards for safety pertaining to industrial robots as they bear weight in the creation of standards for robots outside the factory. The remaining sub-sections discuss additional features of a robot that result in the robot having differing capabilities. The relationship between features and capabilities may be described as follows: the robot is endowed with certain features (ex. auditory, vision or locomotive) which result in the robot having the associated capability.

Capabilities for Safe Interaction with Humans

"An essential component of the duty of care is that a carer must keep their charges safe from physical harm" (Sharkey, 2011, p. 268). In a discussion of robots that may be used for care, I begin with safety to emphasis its significance. The feature of safety renders the robot capable of interacting directly or indirectly with humans in a human environment. Many European initiatives are in place to test the safety standards of the new generation of robots outside the factory. Based on these initiatives, I refer to robots safe for human interaction as human-friendly robots (HFR).

The human-robot interaction may be "hands-on" (Bicchi, 2008, p. 1341) or "hands-off" (Bicchi, 2008, p. 1337). The former refers to robots designed intentionally to interact with humans, a robot to assist with feeding for example, while the latter refers to those robots which may accidentally interact – come into contact with – with a human, for example, a Roomba vacuum cleaner.

Others refer to the interactions between humans and robots according to the flow of information. The interaction is direct if the flow of information is bidirectional meaning "information is communicated between the robot and people in both directions, and the robot and the person are interacting on 'equal footing'" (Thrun, 2004, p. 17). Alternatively, interactions which are considered indirect are referred to as such because "the operator commands the robot, which communicates back to the operator information about its environment, task and behaviour" (Thrun, 2004, p. 17).

The difficulty with building HFRs, or "safe" robots, is the trade-off of speed for accuracy. The challenge these days is to design robots that are safe to interact with humans without having to sacrifice performance criteria. One way to design HFR is using the concept of intrinsic safety: "a robot will be safe to humans no matter what failure, malfunctioning, or even misuse might happen" (Bicchi, 2008, p. 1337). One aspect of intrinsically safe robots is to quantitatively assess the risk of injuries in accidents for comparison with other solutions and for optimization of the robot design. For this, the severity of a potential impact is linked with the statistical probability of causing a certain level of injury. Other methods for designing intrinsically safe robots take the hardware of the robot into consideration to increase their ability to sense objects in their environment or to add protective layers to manipulators (arms) which may come into contact with humans.

Other avenues explored look at introducing mechanical compliance into the design, this means a motor in one area of the robot (i.e. one manipulator) can be decoupled/turned off if an impact has occurred in another area. This design, known as compliant transmission, is thought to diminish performance but this may not be a problem when the robot is used for an entertainment application. In other applications, especially in care contexts, speed and accuracy of task execution are vital.

Standards are already in place for robots used in the factory. The standards pertain to the use of the technology in the workplace but are also robot-specific standards. To be clear, the introduction of robots that can interact with humans in the workplace requires revision of these standards. The International Organization for Standardization (ISO) undertook a revision of standards in 2002 from the original standard for robots in 1992, ISO 10218. One part of this revision is meant to address workplace safety for end-users rather than for manufacturers. This revision in standards allows for new modes of operation between humans and robots in a defined workspace (e.g. simultaneous control of multiple manipulators, mobile robots, collaborative operation). Control reliability no longer relies on hard-wired electromechanical components but rather acknowledges the significance of state of the art software, electronic and network based technology for safety-related soft axis (layers on the hardware of the robot) and space limiting control activity (sensors on the robot to determine and control for their space in an environment). Instead of relying on distance between the human and robot, the new standards recognize that stopping time and distance are more important criteria when robots and humans share a working space.

These standards are particularly relevant when speaking of hands-off robots in an industrial application. Things become more complicated when speaking of hands-on robots which many care robots will be. For hands-on robots, the T-15 committee of the American National Standards Institute (ANSI) is setting safety standards regarding intelligent assist devices. Although these standards cover a wide range of technologies from assistive devices to mobile autonomous robots, they are promising in that they may be translated into policy governing domestic applications of robots. For example, one aspect refers to safety-critical software: under any condition that the robot malfunctions, the entire system will shut down in a safe manner. The standards also indicate dynamic limits which restricts the capabilities of robot design such that a human operator must be able to outrun, overpower or turn off the robot.

One application in which safety is incredibly important is the use of robots for rehabilitation. These robots come in direct physical contact with humans in a variety of ways. For therapy robots, the robot is in direct contact with the disabled patient and the therapist simultaneously. Roboticists in this area must be sure that the robot is designed in such a way that it cannot cause injury by moving a user's limbs outside their range of motion, with too much strength or with too much speed. In addition to this, limits imposed on the robotic apparatus, like additional sensors (referred to as redundant sensors) are used as back-up so if one sensor malfunctions another can identify the problem and shut down if necessary. Outside all this, rehabilitation robots must also be designed to be intrinsically safe; "from the systems perspective, when all else fails, to actively to protect the user, it must be the design itself that makes the robot inherently unable to injure the user" (van der Loos, 2008, p. 1244).

Thus, for care robots capable of hands-off interactions (between the human and robot), intrinsic safety is required in terms of shutting down/off if a problem occurs. Control reliability via software, electronic and networked (if tele-capable) technology for space limiting control activity (stopping time and distance between human and robot). High levels of sensorial apparatus (redundant sensors) for assisting with control reliability. Decoupling motors to ensure if one part of the robot has a problem the whole robot will stop (this will decrease performance but will increase safety). For robots capable of hands-on interactions the above considerations are required as well as safety-related soft axis – the robot is soft to touch.

The remainder of robot capabilities discussed in this chapter are related to the aspect of safety in that each capability renders the robot safer for interaction with a human whether it be the robot's capacity for vision or force feedback.

Robot Vision

Vision for robots may be considered the primary means for sensory input. To achieve vision, visual sensors are required on the outer surface of the robot. This allows the robot to extract a vision from its environment, restore it and enhance the vision (through adjusting pixels) for analysing it. Recognizing the object depends on the stored knowledge of the robot. To do this, however, the robot must be able to accurately define the structure of the objects. Vision allows the robot to find its way about, to analyse chaotic scenes, to recognize faces and/or environments of a human, to detect its own arms and to determine where it is in a given environment (Engelberger, 1989). Laser range finders for vision were shown to enable a robot to create a 2-D map of a nursing home and navigate its way around in the absence of environmental cues (Thrun, 2004). Once the robot has sensed its environment, what it does with this information is left to the control architecture of the robot.

While vision is useful for acquiring information about the robot's environment, it is also useful for acquiring information about the humans present in the robot's environment. Research for detecting people is customary and widespread; "it is common to endow (service) robots with sensors capable of detecting and tracking people" (Schulz, 2003; Pineau, 2003). Robots may be capable of detecting and recognizing gestures (Kahn, 1996), they may track gazes (Heinzmann, 1998) or they may visually perceive head motions, breath expulsions, and/or eye motions (Fong, 2003).

Vision for a care robot is essential for the safety of the humans involved as it provides a means for ensuring the robot can sense its environment to avoid objects. It is also important for the successful completion of tasks (e.g. it can accurately locate an object to be moved or a person to bring an item to). Along the same lines, the feature of vision may endow the robot with more sophisticated vision capabilities. Infra-red vision would allow the robot to detect and locate a human in a dark room (Engelberger, 1989) which would provide a remarkable capability for the care robot, even surpassing the capability of a human care-giver. Researchers are also highly involved in the programming of vision such that the robot is capable of facial recognition (Jain, 2005). Researchers use a variety of techniques to balance light conditions, robot position and human position, but the end result endows the robot with the capabilities for recognizing one face or multiple faces depending on the robot's sophistication and needs.

Above facial recognition, researchers are investigating the potential for the robot to recognize certain emotional states (Kim, 2004; Breazeal 2002; Breazeal, 2008; Mayer, 2010). This is done in a variety of ways like endowing the robot with the capability for perceiving physiological cues and/or bodily movements/gestures as a means for determining emotional state. One avenue for emotional recognition can be seen in the work done at the technical university of Munich's CoTeSys (Cognition for Technical Systems) lab. Using the six universally recognized facial expressions corresponding with emotional states the goal is to make the robot capable of recognizing the human user's emotional state based on its facial expression. This kind of research is meant to facilitate the forming of an empathic bond between the human user and his/her robot (Mayer, 2010).

Auditory Capabilities: Dialogue Management, Voice Synthesis and Voice Recognition

This capability refers to the robot's ability to recognize and understand spoken language as well as the robot's ability to communicate using language. Just like with vision capabilities, the degree to which the robot is capable of any of these tasks may vary depending on the task the robot is required to fulfil. Some robots generate speech but do not understand spoken language (Thrun, 2000), others are capable of understanding spoken language (Bischoff, 2003) or may use keyboard interfaces to communicate using language and bypass speech recognition altogether (Breazeal, 2008).

When the robot is capable of understanding spoken language, it is be referred to as "dialogue management" and consists of a "set of procedures and rules designed to ensure that effective two-way communication is maintained between operator and machine in the face of imperfect (error-prone) communication channels" (Engelberger, 1989, p. 211). If imperfect communication refers to non-recognition of spoken commands, the robot must be endowed with sophisticated software designed for those who suffer from a speech impairment (i.e. stroke patients). If, however, imperfect speech communication refers to a tele-capable robot and the telecommunications link connecting the care-giver and the care-receiver, the robot ought to be endowed with redundant (an additional set of) encoders and decoders to avoid a complete break-down in communication on either side.

Speech as a communication modality is easy to control and may be quite effective for human-robot interactions (Thrun, 2004). There are problems, however, when speech is involved. The number of speakers to recognize and the presence or absence of environmental noise play a role in the success of dialogue management. If the system is "speaker dependant" it will only recognize the voice of a designated speaker (Engelberger, 1989). Programming through demonstration (Friedrich, 1996; Billard, 2008) creates templates which can be matched to spoken words and subsequent utterances will be matched with remarkable accuracy. If, however, the robot must recognize any number of speakers, the software demands increase and the range of vocabulary recognized inevitably decreases to the use of "yes" and "no" in some instances. Environmental noise may also present a problem by masking the sound of the speaker.

One further difficulty with the capability of speech is a misunderstanding, or unrealistic expectations, of the capabilities of the robot. A speaking robot may create a false perception of the robot's level of intelligence, its social capabilities or its overall capabilities (Goetz, 2002; Fong, 2003). One might suggest that maintaining the appearance of the robot as a machine-like artefact may compensate for this tendency.

Auditory capabilities resemble robot vision in the sense that they fulfil requirements for sensory perception and come in a variety of degrees of sophistication. Whether the robot is capable of using this information for the execution of a task or is capable of transferring this information to a human operator who controls the execution of the task is again dependent on the control architecture of the robot. Baseline auditory capabilities, in which the robot can recognize the human voice and match the words spoken with a template held in its memory, may be enough. At other times, sophisticated voice recognition and speech analysis is required (what may be referred to as a social capability).

Mobility/Locomotion

The capability of mobility allows the robot to travel along the x-y planar axis and demands distinct planning and control to achieve this (Kavraki, 2008; Chung, 2008). Mobility can be achieved through a variety of modalities and demands that the robot in motion is programmed for obstacle avoidance (obstacles being material or, in some cases, people) (Minguez, 2008). Mobility is meant to "extend the robotic aide's working volume beyond the desk-top workstation environment. It may also include vertical mobility to facilitate access to floors and shelves" (Engelberger, 1989, p. 213).

Robot mobility is meant to distinguish mobile robots from stationary robots with a fixed platform, like the desktop. The Diet-Assist robot developed at MIT is an example of a stationary social robot which resides in a common room of an individual's house and serves the function of providing support and encouragement to an individual on a diet (Kidd, 2006; Kidd, 2008; Turkle, 2011).

Robot locomotion is different from a robot that is capable of moving an effector or manipulator (arm or hand). For example, the surgical robot daVinci® does not travel as it operates but its robotic arms must be moveable during the course of the surgery. In contrast, examples like In Touch's RP7 robot or the TUG® robot, are meant to travel through the hospital to the patient's bedside and therefore must be capable of locomotion. It follows then that depending on the task for which the care robot is intended, it may or may not be capable of locomotion.

Robots can be made mobile through a variety of modalities; wheels (Campion, 2008), legs (Kajita, 2008), wings (Wood, 2008), or snake-like movements (Hirose, 2009) among others. It should be noted that winged locomotion differs from aerial robotics (Feron, 2008), the latter referring to the development of aerial drones used in military applications (Singer, 2009). Robot locomotion may also be controlled through a variety of modalities; an autonomous mobile robot or a human-operated mobile robot. In many instances, the architecture of the robot is determined according to the chosen means for locomotion.

Mechanics for mobility vary depending on the institution or company designing the robot and the terrain which the robot is expected to move on. Researchers at the Tokyo Institute of Technology are creating a locomotive robot that moves in a snake-like manner (Hirose, 2009). In contrast, researchers at Honda are designing a bi-pedal humanoid robot, ASIMO, to walk like a human using a zero-moment technique (Ng, Thow and Hing, 2009). This technique means the robot equally balances all forces so there is no point at which the robot would lose balance and fall. Additionally, this type of motion requires that the robot be on a smooth surface – not an optimal condition if the robot is to exist in an unstructured environment where these things cannot be accounted for. Other researchers are exploring the use of gravity to propel the "legs" for moving, a technique referred to as "passive dynamics" (Hosoda, 2008). This approach/technique uses little motor power to accomplish walking and is considered a promising, efficient substitute to the zero moment technique used for ASIMO.

Control of the robot's mobility may be human-controlled or autonomous. An example of a human-operated mobile robot is In Touch's RP-7. This robot is aimed at facilitating patient-physician communication when the physician cannot be physically present at the bedside of the patient. The physician, seated at a console in another area of the hospital or in another place entirely, guides the robot through the hallways of the hospital to the patient's bedside. Using a video monitor attached to the mobile autonomous robotic platform, the patient and the physician may communicate directly. In contrast, iRobot's Roomba vacuum cleaner or iRobot's Scooba (pool cleaner) are both mobile robots which operate autonomously; no human manipulation is required to guide the robots locomotion.

In terms of safety, the capability of locomotion requires the use of sensors on the hardware of the robot to indicate if, and when, the robot is approaching an obstacle in its environment (Minguez, 2008). If the mobile robot is autonomous, additional safety considerations must be accounted for – whereas a human-operated mobile robot is less likely to collide with other objects because of the control of the human, an autonomous mobile robot requires redundant (additional) sensors for perceiving their environment. The issue of speed for travelling and stopping in autonomous mobile robots is also relevant. Vision capabilities may be incorporated into autonomous mobile robots to provide the robot with the ability to recognize environmental cues like landscapes or faces. The capability of locomotion, in most cases, will be a pre-requisite if the care robot is intended to fulfil physically demanding care tasks.

Grasping and Manipulating Objects

The earliest robots were praised for their capabilities to grasp and manipulate objects. In fact, these were the only capabilities endowed to traditional industrial robots. For Engelberger, creator of the first industrial robot named Unimate, grasping refers to "the property of a robot that allows specific objects to be selected, positioned and oriented. It is typically associated with 'hands' and includes a variety of functional attributes such as detection of slippage and evaluation of object geometry for stable holding" (Engelberger, 1989, p. 212). Robot grasping is often, although not necessarily, accomplished through robot "hands" (Melchiorri, 2008; Prattichizzo, 2008).

Although reference is made to the term "hands", these are also often referred to as "end effectors" while the arm is referred to as a manipulator. The end effector can take the form of a humanoid hand but can also be a simple gripper consisting of two fingers. These simple grippers can open and close to pick up and let go of a range of small objects. Vacuum grippers are used for heavy lifting of objects with a smooth surface. These types of grippers should not be used for the lifting of humans. Some of the more sophisticated robots have effectors in the form of humanoid hands with up to five fingers (four is more common). Examples include ASIMO, the Shadow Hand or the Schunk hand. The latter two are highly dexterous manipulators (20 degrees of freedom) with an incredible number of tactile sensors. The feedback from the tactile sensors allows the robot to apply the correct amount of pressure so as not to break the egg or light bulb it is holding.

Closely in line with the capability of grasping is that of manipulation (Brock, 2008). Manipulation refers to "the capability to move objects from one place to another while maintaining a correct orientation of the hand and avoiding collisions with stationary objects" (Engelberger, 1989, p. 212). Consequently, manipulation requires knowledge of where the robot is in its environment. Moreover, the robot must have hand-hand coordination to understand when one hand has picked something up and put it back down. In short, we may suggest that the capability of manipulation requires the capability of grasping which presupposes the presence of end effectors (grippers, hands, etc.). This does not, however, imply that if a robot has grippers or hands they will be used for grasping or manipulating. For example, the robot for assisting with elderly care developed at Carnegie Mellon University, named PEARL (Pollack, 2002), has arms but these are used for communication and not for grasping or manoeuvring objects.

Force Feedback and Tactile Sensation

Force feedback and/or tactile sensation features of the robot fall under the umbrella of haptic research (Hannaford, 2008). Although the most significant form of sensory perception for a robot is vision, next in importance may be force feedback and/or tactile sensation. This is especially true in the case of care robots which may have to handle delicate objects like humans, as is the case with robots used for lifting or bathing.

Force feedback refers to the amount of pressure the robot can feel (Cotin, 2000). To accomplish this capability a variety of sensors are required. These "touch sensors" measure pressure applied to various points on the robot end effector, slip detectors to sense loss of a grip and joint-force sensors that measure forces applied by a robot's hand, wrist and arm joints. Placement of these sensors relies on the area which requires feedback. If, for example, the end effector/hand/gripper is meant to perceive the strength with which they are grabbing or manipulating an object, such sensors will be placed on the end effector. These sensors allow the robot to feel whether a grasp is proper for the task and whether there is interference in fitting one object into another (Engelberger, 1989). With this information, the robot may adjust their behaviour accordingly (in the case of an autonomous robot).

Tactile sensation refers to the robot's ability to sense its environment through touch (Kawasaki, 1999). The robot, being able to perceive the variety of textures a surface might have, is then capable of determining whether an object is hard or soft regardless of whether the robot is in the dark or visually obstructed (Engelberger, 1989). Intuitively, one need only imagine the necessity for tactile capabilities in any number of care robots. In the field of surgical robots, tactile sensation and force feedback are of incredible importance. The first surgical robots (Computer Motion's Zeus Telesurgical System and the daVinci® robotic system) were not endowed with such capabilities and thus surgeon's had to learn new skills in order to perceive the surgical field properly. The latest in surgical robots, namely the Amadeus Composer, aim to incorporate haptics into the architecture of the robot. Those that are used for lifting, bathing, or even feeding patients will all require haptic capabilities (force feedback and/or tactile sensation) to some extent.

Social Communication/Capabilities

Social communication may be considered a feature of a robot which is facilitated by a range of social capabilities. Social capabilities of the robot refer to the ways in which a robot can communicate with a human user in an engaging, interpersonal, social manner, essentially in a more human-like manner. At this point in time social robots are considered a class on their own because of the technical requirements to program them accordingly. As technology progresses, it is believed that the function of social interaction will be a feature added to the existing technical framework of other robots like the domestic robot Roomba or even industrial robots.

The field of social robotics began around the 1940–1950s by William Grey Walter and has been developed extensively in the 1990s by researchers like Kerstin Dautenhahn (University of Hertfordshire) and Cynthia Breazeal (MIT). The work of Dautenhahn has focused on creating a kind of robot etiquette outlining proxemic cues to be programmed in the robot to ensure the robot complies with the preferences of humans for successful interaction (Dautenhahn, 2007). Dautenhahn is also engaged in extensive work investigating robots endowed with social capabilities used as teaching tools for children with autism (Dautenhahn, 2004; Dautenhahn, 2003).

The work of Cynthia Breazeal has focused on different aspects. Breazeal aims to create robots that can interact over a long period of time with humans and do so in a meaningful way: "social or sociable robots are designed to engage people in an interpersonal manner, often as partners, in order to achieve social or emotional goals" (Breazeal, 2008, p. 1349). The domain of social robotics is motivated by questions like: "how to design for a successful long-term relationship where the robot remains appealing and provides consistent benefit to people over weeks, months and even years"? (Breazeal, 2008, p. 1350).

Social capabilities are distinct from physical capabilities in terms of the ends to which they serve: social capabilities are intended to establish and/or maintain a relationship while physical capabilities are intended to fulfil some kind of physical labour. "The benefit that social robots provide people extends far beyond strict task performing utility to include educational, health and therapeutic, domestic, social and emotional goals and more" (Breazeal, 2008, p. 1350).

These robots interact with humans in a social way, meaning they communicate (visually, auditorily or verbally) with humans beyond indicating the initiation or completion of a task. "Social robots use a variety of modalities to communicate from whole-body motion, proxemics (i.e. interpersonal distance), gestures, facial expressions, gaze behaviour, head orientation, linguistic or emotive vocalization, touch-based communication, and an assortment of display technologies" (Breazeal, 2008, p. 1350). The interpersonal manner in which these robots are meant to engage people relies on verbal as well as non-verbal cues (referred to as paralinguistic cues). The robot must be able to perceive this information, interpret it accurately and respond appropriately. The issue of interpretation is quite complex due to the range of human behaviour and communication modalities and thus social robots are considered among the more sophisticated robots of today. This type of interaction with a human user presupposes that the robot is safe to interact with humans in either a hands-off or a hands-on form.

Social robots may be used for enjoyment, learning, therapy or for personal growth. Philips ICat, for example, is used to understand human-robot interactions in order to program future robots for greater user acceptability. In therapeutic instances, social robots are used to interact with children with autism. The work of Dautenhahn shows how social robots can be used as a tool for teaching children skills of interaction. Robots with social capabilities are also used as diet assists; individuals wanting to lose weight use the robot to help motivate, encourage and keep track of progress or lack thereof. The goal is to foster a meaningful bond between the human and the robot in order to achieve weight loss goals with greater success.

Aside from these instances, sociable robots, or the capabilities of social robots, are also used in service applications. For example, a robot to greet people as they enter an office, a hospital, an airport or an elderly home (Vongsoasup, ND). In this case, the robot is not intended to form a long term bond with users; however, the same capabilities for such an initiative are used to facilitate successful human-robot interactions to meet a service goal (Shieh, 2007; Fong, 2003). In these scenarios the robot might understand and produce natural language, or may understand and produce proxemic cues. Sociable robots used in these "service situations" fulfil a short term functional goal rather than social robots used to meet the long term emotional needs of persons.

In terms of care robots, communication in the interpersonal manner for which they are designed, seems quite appropriate for the interaction between a care-receiver and a care robot. This may make the care-receiver more comfortable having the care robot in their personal space, whether it be their home, hospital or nursing home room. Additionally, if the robot is able to communicate in a social manner (whether verbally or via paralinguistic cues) it may make the care-receiver feel more at ease having a care robot involved in a care practice to begin with. One might suggest then, that a care robot intended to provide care for the care-receiver directly ought to be equipped with social capabilities to the extent that they may sense and perceive a range of verbal and non-verbal communication paradigms and may communicate with a human user in a similar fashion. This may also ensure the robot acts politely according to cultural standards, a robot etiquette if you will.

One might also suggest that communicating in an empathic way seems quite appropriate as a capability for a care robot. Such a capability may be facilitated through facial recognition or emotion recognition (as is studied at CoTeSys in Munich). For the delicate and sensitive tasks in care that require a caring, or empathic, disposition we may even go as far as suggesting that a care robot *ought* to be endowed with such features to ensure the capability of forming an empathic relationship. However, one might also alternatively question whether this is not simply an affirmation that a human care giver ought to be present. Taking a second glance, is it wise to have the robot capable of eliciting this type of communication? If the care-receiver believes the robot to be empathetic could it not interfere with the care-receiver communicating, or wanting to communicate, with the human care-giver? If the robot were endowed with this capability, is it possible that a human care-receiver will find it easier to communicate with the robot than with a human care-giver (Pransky et al., 2004). I will address these questions in the next chapter dealing specifically with care robots and their capabilities.

Appearance

The appearance of a robot is not necessarily a capability but I refer to it as a feature which allows me to discuss it in this chapter. It is one of the most important features as it plays a large role in the expectations users have towards the robot as well as their comfort level with, and acceptance of (or lack thereof), the robot. Reasonable expectations are required for a functioning interaction between human and robot: if a human expects the robot to behave a certain way and it does not then the interaction will fall apart. The same holds for comfort level with the robot and acceptance of the robot. Studies show there are limits to the robot's appearance in order to make humans more comfortable when interacting. Both expectations and comfort level add to the acceptance of a robot which is incredibly important when we consider that robots are now entering into our personal spaces (e.g. homes and nursing homes).

Although many engineers believe it is favourable for the robot to have a humanoid appearance (and have been designing according to this assumption), many believe this confuses the expectations of users. According to a survey done by Fong et al. (2003) participants claimed that a robot's appearance ought to correspond with their intended tasks. If the robot is meant for cleaning it need not appear human-like but may appear as a machine, like the Roomba.

Aside from humanoid robots another category of human-like robots is that of androids. In terms of their appearance, they are designed to resemble a human with as much detail as possible, including skin, teeth, eyes, etc. A US company called Hanson Robotics is currently involved in the design of android robots for entertainment purposes, their most well-known android being the Einstein head (Oh, 2006). Even today, in 2014, androids are not used in service applications of any kind but are used instead to test human reactions to robots resembling humans to varying degrees (Minato, 2004; Sakamoto, 2007). The reactions of humans are studied and described using the uncanny valley theory (Mori, 1970). The "uncanny valley" is the phenomenon whereby humans have feelings of disgust when they view a robot that looks nearly, but not quite, like a human. The robot appears creepy in its resemblance because something is just not right. It is possible to relieve this feeling of disgust by adjusting the appearance to either increase or decrease the robot's resemblance with a human (Mori, 1970). This dip in acceptance is what is referred to as the "uncanny valley".

The appearance of today's functioning robots are mostly machine-like, for example: the RP-7 robot, the TUG® or HelpMate™ delivery robots, the daVinci® surgical system, and the Sanyo bathtub to name a few from the healthcare domain. Robots can also be creature-like resembling a known creature like Paro, the white baby seal with fur, or ambiguous like Keepon the small dancing robot that looks like two yellow tennis balls stacked one on top of the other. Keepon has a simple face and expresses itself by squashing or stretching its body.

For care robots there is once again no defining appearance that signals the robot is a care robot. They may have any kind of appearance. Interestingly, selecting the robot's appearance will depend greatly on its capabilities and use and it will be current and future user studies that will indicate the preferred appearance for the care robot.

Modes of Robot Control

The discussion up until this point has addressed the appearance and capabilities a robot may have which are sensory capabilities of the robot. I now turn your attention to how these sensory mechanisms are controlled. Using the information from the robot's sensing capabilities, an action or task must then be accomplished. There are roughly two options for executing the task; human-operated control or autonomous control. In short, human-operated means a human is responsible for executing a task once given the information provided through the robot's sensing capabilities/sensors. In contrast, autonomous control means that the robot will accomplish a task given the information it has acquired through its own sensing capabilities/sensors. That being said it should be noted that there is a range within these two general options: a robot could be partially autonomous for one portion of a task and human-operated for another.

Human-Operated Robots

Human-operated robots, as a type of robot, have the feature that they must be operated on, or controlled by, a human in order to accomplish their required task or action. Human-operated robots represent the first robots commercially used in the factory. One configuration of human-operated robots is referred to as "master–slave" (the human operator as master and the robot as slave); the movements or commands of the human are translated into movements made by the robot.

For hands-on human-robot interaction in industrial applications, the robots are often referred to as "cobots". These are "collaborative robots" designed to relieve humans from fatigue or stress and to prevent injuries; "cobots presume a division of control between human and robot, with a robot perhaps supporting a payload and allowing a human to guide it" (Bicchi, 2008, p. 1345). In this scenario, the operator is in direct physical contact with the payload. This description also describes exoskeletons used for rehabilitation purposes (Hayashi, 2005); "exoskeletons are also controlled by a human operator, leaving all planning and high-level challenges to the user" (Niemeyer, 2008, p. 741). Again, the user is in direct contact with both the robot and the payload.

Within the class of human-operated robots is a subclass known as telerobots whose infrastructure is designed so that a human operator can control the movement of the robot with the added condition that the human operator is at a distance (Niemeyer, 2008; Mitra, 2008). Again, all planning and cognitive decisions are made by the human user and the robot is used strictly for mechanical completion of a task. The use of "tele" (derived from the Greek word for distant) presumes a geographical separation between the user and the environment in which the task is being performed. The inaccessibility of the environment may be for any number of reasons; the user cannot or will not physically reach the environment (as in robots used to search pipes for gas leakages), the environment is dangerous (as in search and rescue robots, robots used in marine or extra-terrestrial environments), the environment needs to be scaled to be made accessible (as in surgical robots).

The physical distance between the user and the robot varies depending on the application (e.g. for surgical robots the surgeon is often in the same room, for robots in space or underwater the distance is much greater). In most cases there are two sites to speak of; the local site with the human operator and the remote site with the robot. For the information to travel from one side to another, the two sites must be connected. Traditionally this was done through the use of cables; however, recently computer networks have made it possible to transmit this information from one side to another using a telecommunication system (Rayman, 2006), a dedicated network, or a satellite in some instances (Rayman, 2007).

In human-operated robots control of the robot may occur through one of three architectures; direct control, shared control, or supervisory control (Niemeyer, 2008). Direct control assumes no autonomy or intelligence on the part of the robot, thus, all the motions of the robot are directly controlled by the user (e.g. the Hybrid Assistive Limb [HAL] exoskeleton). Shared control refers to a sharing between local and remote sites whereby the human operator decides what to do and how to act while the robot can autonomously refine the command for the environment. This is the case for the daVinci® surgical platform; the surgeon performs its movements which the robot autonomously scales down to the appropriate size for the surgical field. Supervisory control is described as analogous with supervising a subordinate staff member whereby the supervisor is responsible for giving orders to the subordinate but in turn receives summary information (ibid.). It may be suggested that autonomous robots evolved from the design of supervisory control robots (Haselager, 2005).

With these three levels of human-operated robots the human operator may take over control of the task at any time. Accordingly, all three of these models appear appropriate for the use of care robots. Furthermore, a degree of control by a human seems an intuitive requirement when the physical safety of the care-receiver (or care giver for that matter) is at stake.

Robot Autonomy and Intelligence

In contrast to robots controlled directly through the input of a human user are a class of robots called "autonomous" robots. Autonomy of the robot does not refer to the robot's capability to sense, nor does it refer to the systems which command the robot's motor responses but rather, the robot's capability of interpreting the sensory input in order to make a situational judgement for action (Engelberger, 1989, p. 99).

The issue of robot autonomy is problematic because of the varied conceptions one may have of the concept of autonomy. While philosophers approach autonomy in terms of the freedom to, and reasons why one, act(s) in a certain way, roboticists approach autonomy from the question of how the robot fulfils its task (with or without assistance or supervision); "within robotics, the increase in autonomy of a system is related to the reduction of on-line supervision and intervention of the operator, programmer or designer in relation to the robot's operations in a changing environment" (Haselager, 2005, p. 518). This engineering interpretation of autonomy says nothing of the robot's freedom to choose its actions but captures the idea of how the robot acts. For the purposes of this book I use this idea to discuss autonomy and refer to autonomy in terms of the robot's capability to fulfil a task without real-time manipulation from a human operator.

Restricting robot autonomy to this conception, however, does not capture the nuanced way it is currently discussed. For Engelberger, "autonomous planning is performed by the machine when sensed data are operated on by application programs with the result that the machine makes navigating (or equivalent) decisions" (Engelberger, 1989, p. 211). These decisions do not require human interaction but are, on the robotic side, "subject to human supervision and veto" (ibid.). Later visions of autonomous robots claim that the robot may "operate under all reasonable conditions without recourse to an outside designer, operator or controller while handling unpredictable events in an environment or niche" (Franklin, 1997).

These two definitions of robot autonomy maintain that the robot is acting according to a pre-programmed set of rules, and the robot is capable of planning their actions without referring to a human operator (or designer or controller) during execution of the task. What the first definition (Engelberger, 1989) allows for is a human to observe and veto an action. What the second definition adds is the capability of the autonomous robot to fulfil its task in an environment in which it has not been trained, and/or one that is unpredictable.

This is the main difference between industrial robots and all other robots used outside the factory – for the latter it is impossible to program all aspects of the environment that the robot will encounter making it unpredictable and unreliable. With this in mind, the more recent conceptions of robot autonomy now include the criterion of adapting to their environment; "autonomy refers to a robot's ability to accommodate variations in its environment" (Thrun, 2004, p. 14). Other definitions claim autonomy also includes the capability of the robot to fulfil its task within time constraints and with the added component of potential interference by others: robots must: "operate in highly variable, uncertain, and time-changing environments; meet real-time constraints to work properly; interact with other agents, both humans and other machines" (Bensalem, 2009, p. 67).

In short, we may observe that the feature of robot autonomy renders the robot capable of different things depending on the conception of autonomy one holds. Without prescribing what an autonomous robot should refer to, we may suggest that it has the following properties: it can perform its pre-determined task in an unpredictable environment without consulting an outside source for assistance; the task is performed under time constraints in an unstructured and/or dynamic environment; if humans are present in its working environment, depending on the robot's function, the robot will be capable of interacting with humans in a hands-off and/or hands-on manner and as such the associated safety considerations are required.

For Thrun (Thrun, 2004), a robot's autonomy has both types and degrees. Types of autonomy refer to a robot being capable of making decisions about its environment or of making decisions about a human's behaviour. Autonomously navigating through an unpredictable, unstructured or hazardous environment presupposes the robot be capable of acquiring environmental models. Alternatively, a robot may be capable of detecting people and their behaviours to autonomously accommodate for them. This means that when referring to a robot's autonomous capabilities, it is important to clarify in which way the robot is autonomous; adapting to its environment or adapting to its user. The degree to which the robot is autonomous has to do with the sophistication of the programming and the amount of tasks a robot may be capable of fulfilling in an autonomous manner. The latest in autonomous robot research refers to robots capable of learning. A learning robot may be considered the most sophisticated form of autonomy to date.

Robot learning

Robot learning may be used to refer to a feature of a robot – the robot can adapt by changing its behaviour based on its previous experience (Franklin, 1997) – or, to the way in which the robot is programmed – learning by demonstration (Friedrich, 1996; Billard, 2008), mimicking (Mayer, 2010), or reinforcement (Billard, 2008; Santoro, 2008). The concept of robot learning invariably increases the degree of autonomy the robot has and increases the success with which the robot will manoeuvre in a new, unknown environment. Of course this way of acting and interacting inevitably invites the concern that a robot is then free to choose a certain course of action. One may wonder how we can ever predict what the robot will do? And if we can't predict what the robot will do how can we ever ensure that it is safe?

With respect to programming robots by learning, it is thought that robots learn general rules from their experience in order to meet task assignments in highly variable environments (meaning human environments) (Santoro, 2008). There are many ways in which roboticists are exploring how to program learning into the robot. The "Child-robot" developed in Suita, Japan, is said to develop social skills by interacting with humans and watching their facial expressions, mimicking a mother-child relationship (Minato, 2007). The aim of the creators at Osaka University in Japan is to develop this robot to think like a baby, meaning the robot will be able to evaluate facial expressions and cluster them into basic categories like "happy" or "sad". These robots are considered the predecessors of more advanced social robots.

The kind of learning of the robot will depend on its context and use. Learning robots will not always have to be safe for human interaction because they may be applied in military, surveillance, or search and rescue applications.

Robot cognition

Another example of sophisticated robot autonomy is seen in the work done at the Technical University of Munich. In the CoTeSys lab, roboticists are investigating the ability to program a kind of cognition into the architecture of the robot (Zaeh et al., 2010; Buss et al., 2010; Tenorth 2010; Tenorth 2012; Tenorth, 2011). There the robot is expected to know: what it is doing and why; what it sees and saw; what it is capable of doing and what it is not capable of doing; and predicting the consequences of its actions based on such reasoning. The manner in which this is accomplished is through the programming of semantic links between action codes (Kunze et al., 2011).

To give an example, the CoTeSys lab works with two robots to make pancakes (Beetz et al., 2011). One robot is mobile, and used to gather the ingredients, and the other is stationary, used to pour the mix on the pan and flip the pancakes. The robots are given verbal commands and respond to the commands verbally. If one were to ask the stationary robot to fetch the milk from the fridge, the robot would respond that it is not capable of moving and thus cannot fulfil the request. This robot is thought to be cognizant of its own capabilities and the limitations thereof. The idea is that such capabilities allow for more human-like interactions as well as building trust between the human and robot counterparts when the robot can account for its own actions. One might wonder whether the robot is capable of being responsible if in fact the robot is cognizant (or aware) of its own actions and the reasoning for such actions. This will be the topic of discussion later on.

Artificial Intelligence

Closely in line with discussions of sophisticated forms of robot autonomy, and machine learning in particular, is a discussion of robot intelligence, also referred to as Artificial Intelligence (AI). It is beyond the scope of this book to provide a comprehensive overview of the field of AI; however, for the purposes of this book, it is important to introduce this field given its role in the development of current and future (care) robots. The field of AI is concerned with the study of intelligent beings, just as philosophy and/or psychology is; however, AI strives to build intelligent entities – hence, artificial intelligence.

The study of intelligence demands the question of how one defines intelligence. The Turing Test (Turing, 1950), proposed by Alan Turing was intended to provide an explicit, concrete, definition of intelligence: "the ability to achieve human-level performance in all cognitive tasks, sufficient to fool an interrogator" (Russell, 1995). In order for the computer to be "intelligent", it was/is assumed that the computer "needs to possess the following capabilities: natural language processing, knowledge representation, automated reasoning and machine learning" (Russell, 1995, p. 5).

There are a range of approaches within the field of AI for programming such capabilities: the cognitive modelling approach, the laws of thought approach and the rational agent approach (Russell, 1995, p. 6). This high level intelligence renders the robot capable of a sophisticated kind of reasoning. Such reasoning has to do with the way in which information input into the system is processed before giving output. The robot may be mobile and as such taking in information pertaining to their environment. As we saw in the case of mobile robots, path planning and object avoidance are key issues. Using AI, autonomous mobile robots are capable of such tasks without the direct input of a human user.

Another recent avenue that AI is taking is that of Affective Computing (AC) (Picard, 2000). The goal is to endow computers with the capability to perceive emotions of human users in order to improve human-computer interactions (i.e. make these interactions more intuitive) (Oosterhof, 2005). The main questions of AC deal with how a computer can detect emotions and whether a computer should itself have "emotions" (which of course presupposes the more philosophical question of whether it is possible for computers to have emotions). Recognizing emotions requires that the computer (or robot) have sensors that go beyond those already described. The robot's vision capabilities must be sophisticated in the sense that it can detect facial cues (and not just faces). The robot's auditory capabilities may also allow the robot to perceive emotional information about the user (e.g. the volume the user uses to speak). The robot could also perceive physiological changes in the user as a means for emotion perception. Additional kinds of non-verbal cues the robot has to perceive refer to bodily cues like gestures and or proxemics (i.e. spatial distances).

The field of AC is of great significance in a discussion of care robots. As we have seen in Chapter 2, good care is dependent on the nurse's ability to perceive the emotions of patients (i.e. attentiveness of the nurse) and to tailor their behaviour accordingly (i.e. competence of the nurse). The question then becomes whether or not robots used in care must be endowed with such capabilities given their mere presence in a care context.

With an understanding of the variety of robot capabilities, features, appearances, and modes of control the aim now is to provide a definition of a care robot, current examples of care robots for analysis, and to project future capabilities of a care robot based on the care analysis of the previous chapter. Exploring care robots in terms of their capabilities and definition is the subject of the next chapter.

Chapter 4
What is a Care Robot?

Introduction

This chapter aims to address the technology of care robots. To do this, I outline what a care robot is, the functions it serves, and the technical capabilities a care robot may possess now and in the foreseeable future. This chapter also aims to address the definition of care robots in conceptual terms – how can we call a care robot as such, especially now that we understand the many images that *care* conjures.

Defining a Care Robot

There is not one capability, appearance or function that is exclusive to a care robot. It may have any number or combination of capabilities and appearances. For this reason, I use the concept of interpretive flexibility (Howcroft, 2004) to discuss their classification. According to interpretive flexibility, a care robot is classified based on: its context of use, the function for which it is used, and the user. Thus, one robot may be referred to as a care robot when used by a nurse in a hospital setting to lift patients but the same robot may also be classified as an industrial robot when used in a factory by factory workers for lifting heavy objects (e.g. an exoskeleton).

When defining a care robot, there are a few definitions currently available: "Carebots are robots designed for use in home, hospital, or other settings to assist in, support, or provide care for the sick, disabled, young, elderly or otherwise vulnerable persons" (Vallor, 2011). For others, defining a care robot has to do with the anticipated roles it is expected to fulfil. Sharkey and Sharkey (2011) discuss care robots specific for elderly persons and list three main uses of such robots: to assist the elderly, and/or their carers in daily tasks; to help monitor their behaviour and health; and to provide companionship. Alternatively, care giving robots may be distinguished between an "affective robot which refers to if/when a robot is supposed to be a friend or companion to a human being as opposed to a utilitarian robot which refers to if/when a robot is used in a similar manner to a tool or instrument" (Shaw-Garlock, 2009, p. 250).

For the purposes of this work, I align myself closely with the first definition of a care robot which does not specify demographic of use. My definition of a care robot takes into account the care ethics perspective, thus, I claim that care robots may be defined as any robot used in a care practice to meet care needs, used by either or both the care-provider or the care-receiver directly, and used in a care context like the hospital, nursing home, hospice or home setting. With this definition I aim to make clear that a care robot is one which will be integrated into a care practice and consequently is integrated into the therapeutic relationship between care-giver and care-receiver.

As was evident in Chapter 2, the expression of care values is dependent upon care practices and the relationship between care-giver and care-receiver. In care contexts, such a relationship is referred to as a therapeutic one and is established through care practices like bathing. This is not to say that a care-receiver using a robot on their own (in a hospital or home setting) is not using a care robot; the robot has been provided to them by a care-giver through a care institution. Those robots which are commercially bought and used in home settings, I classify as domestic robots, and they are distinct from care robots. They may serve care purposes but by virtue of their acquirement, and without being integrated into the care relationship, they cannot be referred to as care robots. With this in mind, I hold that the care ethics perspective, and its emphasis on relationships, is prominent in the very definition of a care robot.

In current academic discourse, care robots are thought of as a class of robots to be used by nurses, as opposed to surgeons. This is due to the fact that care robots are thought to assist with physical care tasks, activities of daily living (ADLs) such as lifting, bathing and feeding, all of which are fall within the realm of the nurse's responsibilities. Although there are scholars claiming that surgical robots should be categorized separately from care robots given that one is used by the physician and the other by the nurse (Veruggio, 2006; Vallor, 2011), I claim that surgical robots too ought to be considered care robots and consequently should also be evaluated in the same way (i.e. according to the framework proposed in this book).

I argue this for multiple reasons. First, the surgical robot is one that is integrated into a therapeutic relationship. Second, the use of the robot not only changes the way in which the nurse assists in surgery (nurses must undergo training with the robot) but changes also the ways in which nurses care for the patient post-operatively. Remember from Chapter 2 that when we consider care at the level of the institution, or care as a process, we must recognize the linkages between practices. Consequently, this robot, integrated into the healthcare tradition, although used predominantly by the surgeon, also changes the way in which care is practised by the nurse.

Examples of current care robots include: the Secom MySpoon automatic feeding robot; the Sanyo electric bathtub robot that automatically washes and rinses; Mitsubishi's Wakamaru robot for monitoring, delivering messages, and reminding about medicine, and Riken's RI-MAN robot that can "pick up and carry people, follow simple voice commands, and even answer them" (Sharkey, 2011, p. 267). The RI-MAN robot has now been replaced with the RIBA robot which has an animal-like appearance as opposed to RI-MAN's humanoid appearance.

Other examples include Paro, the baby seal for companionship; RP-7, the mobile robot for patient contact with a physician not geographically present in the hospital; the daVinci® surgical robot; Titan's new Amadeus Composer surgical robot; the medication reminding robot developed by Susan and Michael Anderson (Anderson, 2010) (University of Connecticut and University of Hartford respectively); the TUG® and Helpmate™ robots used for the delivery of sheets, medication and/or food tray removal in the hospital. In more sophisticated care institutions, robots are used for security and monitoring purposes; "In the high-tech retirement home run by Matsushita Electrics, robot teddy bears watch over elderly residents, monitoring their response time to spoken questions, and recording how long they take to perform certain tasks. These robots can also alert staff to unexpected changes" (Lytle, 2002).

As we can see there is no capability or appearance exclusive to all care robots. They may be used in a variety of contexts from the hospital to the nursing home, hospice or home setting and they may fulfil a wide range of activities. What they all have in common is their integration into the therapeutic relationship.

Care Robots, Social Robots and Companionship

Given the distinction between affective and utilitarian purposes (Shaw-Garlock, 2009) made above, a further clarification must be made with respect to my definition of a care robot: a distinction between social robots and care robots. As was seen in Chapter 2, care practices for ADLs were done in a way that also met the social and/or emotional needs of the care-receiver: it is impossible to definitively separate the meeting of needs along the dimension of physical care and social care. Because of this, can we claim that a social robot may in fact be a care robot when, by definition, social robots meet social needs?

I suggest that it is not possible to qualify a social robot as a care robot. Care robots are intended to meet the care needs of individuals. Such needs are fulfilled through care practices and rely on the therapeutic relationship. Thus, the relationship or bond between care-giver or care-receiver and the care robot is not an end in itself. Rather, is it a means to the end of meeting care needs. Alternatively, with social robots, the end goal is the establishment of a bond between robot and the human user; to make the robot a companion. Thus, the end of the social interaction between human and robot is the formation of a relationship rather than some other end; the social robot is engaged in social practices rather than care practices. It is possible to suggest that a robot be designed to engage in a social way in order to meet other needs, remember the diet assist robot. Here I argue that the robot is not in fact a social robot but is a care robot with social capabilities.

Added to this is the kind of relationship that is formed in the two interactions. The goal of the relationship between a social robot and human user is that of a companion, of companionship (Breazeal, 2004). It is possible to suggest that a specific type of care need is that of companionship but the question remains whether or not this is the kind of need met by the nurse. In response, the relationship between care provider and care-receiver is a therapeutic one, explicitly differentiated from one of companionship: "the therapeutic relationship differs from a social relationship or friendship in that the needs of the client always come first. The nurse is in a privileged position because of the trust the client puts in the nurse and because of the power imbalance" (Nurses of Ontario, 1999a, p. 8). The differences have to do with the explicit recognition of the asymmetry in power and the ends which the relationship serves (care needs along a broad spectrum).

Moreover, the institutional context within which the relationship is established plays a role in creating the boundaries of the therapeutic relationship. It should be noted that as healthcare and nursing practices are changing so too is the context of care along with the roles of the nurse. Care is being provided in the community and in certain instances in the homes of patients which shifts the activities of the nurse and has the potential to blur the boundaries of the relationship:

> The nurse may be taking on a stronger counselling role with clients and/or focusing or concentrating on psycho-social issues. The nurse may need to clarify the role for him/herself and explain that role to the clients who may be expecting a more traditional role from the nurse … In some instances the role of the nurse can include teaching clients how to grocery shop or do banking, or a community nurse may be involved in planning meetings in the community. The nurse needs to be clear with the client that this activity is part of the nursing role and not an extra activity outside of that role. (Nurses of Ontario, 1999a, pp. 10–11)

Another manner in which the boundaries of this relationship are made clear is evident in its termination: "at the beginning of the relationship, the nurse establishes with the client, family and health team an estimated period of time that the relationship will last. The health-related goals and need of the client determine when the relationship will end" (Nurses of Ontario, 1999a, p. 11).

From all of this it is quite clear that the boundaries of the therapeutic nurse-patient/client relationship are managed through the actions and attitudes of the nurse regardless of the context within which care is being provided. We can see that there are many ways in which a therapeutic relationship differs from one of companionship and as such a social robot, defined by its initiative to foster companionship, cannot be considered a care robot.

This is not to say that care robots will not or cannot have social capabilities. Take the example of Cody, the diet assist robot developed at MIT, it is a robot with the end of keeping the user motivated towards their goal of weight loss and the bond formed is an integral component of this. Cody is a care robot with social capabilities. Its end-goal is weight loss and not strictly becoming a companion.

Discussing the boundaries of the therapeutic relationship reminds us of the many attributes of the nurse and the many values expressed in, and through, a care practice. This brings us to an important question: how do we translate values and human attributes into capabilities of a care robot?

Designing a Care Robot According to the Values in Care

If we consider that a care robot will be used to fully or partially replace a human care provider in a given care practice then the robot ought to be evaluated along the same lines as the human care-giver. Not only this, but for the prospective design of care robots, designers should aim to endow the care robot with those capabilities and attributes that make a human care-giver a good one.

In terms of design, for Engelberger, "the better a robotic aid can match tasks with capabilities, the more useful it will become" (Engelberger, 1989, p. 211). Thus, one must understand the human capability in order to program the robot accordingly. Table 4.1 provides an outline matching human capabilities to robot capabilities. I have chosen to list the human capabilities according to the necessary elements indicated by care ethicist Joan Tronto: attentiveness, responsibility, competence, and responsiveness. As we may recall from the previous chapter, these elements provide the criteria for ethically evaluating whether or not care is good care. What's more, they may loosely be considered attributes of the good care-giver and as such provide a starting point for the evaluation of a care robot.

According to Table 4.1, one might suggest that in order for a care robot to be considered attentive, it must be able to, at the very least recognize the face of the user (care-giver and/or care-receiver). When discussing such a task, however, using examples we can see how a difference in the context, practice and/or actors changes whether or not the robot capability conforms with the requirements of the care practice or if it presents itself to be ethically problematic.

Table 4.1 The moral elements of a care practice (Tronto, 1993) aligned with corresponding robot capabilities

Moral element	Translated into human capability	Translating human capability into robot capability
Attentiveness	Capability and capacity of recognizing the dynamic needs of a patient.	Robot Vision; Facial recognition, Emotion perception and recognition.
Responsibility	Closely aligned with trust; requires an understanding of what one is doing and why. Capability of identifying adequate response to needs and delegation to meet them. Presumes individual will be held accountable and liable in the case that something goes wrong.	Emotion recognition as a means to establish trust (mimicry). Robot knows what it's doing, how and why it's doing it . Knows what it sees, saw and can see. Can predict the consequences of its actions. Know what it can and cannot do. Can acquire new knowledge (learning robots/algorithms).
Competence	Capability and capacity of executing an action to fulfil the identified needs in a skilled manner.	Safety, efficiency and quality of task execution (speed of robot, stopping distance, emergency shut off/power down, materials used for robot). Force feedback and tactile perception.
Responsiveness	Capacity and capability to engage with the care-receiver regarding the meeting of their needs (can be physiological, verbal or other cues given by the patient to the care-giver).	Multi-modal communication platforms: verbal and non-verbal/paralinguistic communication paradigms, hand gestures, proxemics, head gestures, lights, eye gaze, facial expressions, force feedback and tactile sensation.

As an example, let us discuss the element of attentiveness and endowing the robot with facial recognition (so that the robot recognizes the patient) in order to say that it meets even a portion of the requirement of attentiveness. What happens when we observe this capability in more than one practice, more than one context and more than one robot type? In a practice like lifting, if the robot were autonomous, a capability like facial recognition seems appropriate to guarantee that the robot is aware of the individual patient it is lifting and further that the robot is privy to other information about the patient which it can access once it has the patient's facial information (the context is the hospital or nursing home where the robot would have more than one patient to lift).

What if we take the same practice, in the same context but we look at a human-operated robot, the exoskeleton. First, is facial recognition required if the robot is essentially acting as an assistant to the nurse? Second, perhaps attentiveness in this practice refers not to the robot's need to recognize the patient but the robot could enhance the attentiveness of the care provider –

attentiveness as defined for the network and for the practice. If that were true then the robot need not have facial recognition capabilities at all. Switching contexts, if the autonomous robot was in the home where only one person is being lifted, there is no need for facial recognition nor the information (or attentiveness) acquired through such facial recognition.

More than Values is Needed for Design

This example introduces the idea that a robot need not be endowed with certain capabilities if their role does not demand it. Would a surgical robot need to be capable of facial recognition of the patient? If the robot were autonomous one wouldn't hesitate to agree; however, when the robot is human-operated the element of attentiveness is always the human's role and responsibility. This is not to say that the robot could not enhance the attentiveness as it is distributed throughout the network. Rather, the robot is delegated a role in which its responsibility is to enhance the capabilities for attentiveness of the human care-giver rather than take-on the role of attentive care-giver.

This also means that attentiveness for a particular practice must be understood as it differs between practices. Attentiveness for lifting refers to the mechanics of lifting as well as its affiliation with the establishment of a relationship between care provider and care-receiver, and the overall process of care (as was discussed in chapter two). Alternatively, attentiveness as it pertains to surgical intervention is specific for the type of surgery but is nonetheless defined by the surgeon's perception of physiological cues and bodily responses to surgical intervention rather than in the establishment of a bond.

The same holds for the other moral elements: competence, responsibility and reciprocity must be understood according to individual care practices. Competence and its relationship to safety differs between the practices of lifting and of surgery. Safety during surgery requires a sterile environment along with the resources for the procedure. Safety for lifting requires strength and force feedback not to drop the patient.

The interpretation of the moral elements themselves are dependent on the context, the practice, and the actors involved. Isolating the moral elements from these components is logically incoherent given that the meaning and prioritization of moral elements is dependent on all the other factors. In short, we are talking about an entire network of human and non-human actors which all contribute in some way to the manifestation of values. Thus, all are delegated a role and responsibility. The care robot is not evaluated against the criterion of a human care-giver necessarily, but is evaluated according to a broader set of criteria, one that encompasses the role and impact it may bear on the network of actors and the ethical nature of the care practice.

When it comes to design recommendations for robots Sharkey and Sharkey also stress the need for taking the larger picture into analysis; who is the demographic, what kind of information will be collected, can users opt out, what happens to the information if they have given it and want to take it back, etc.? I conclude from this as well as the above analysis of translating the moral elements into robot capabilities that design recommendations must take into account not only the values at stake but an in-depth study of the practice into which the robot is stepping. Such a study is dependent on the context, the actors, the distribution of roles and responsibilities, and the resulting values manifest through the interactions and actions of all actors. Thus, the framework for evaluating a care robot must incorporate all of these components in order to evaluate the robot's capabilities in ethical terms. With this, the stage is set for the various components of the framework. In the next chapter I outline the framework for the ethical evaluation and design of care robots along with its method for use.

Chapter 5
A Framework for Evaluating the Design of Care Robots

Introduction

The overarching question that this work has set out to address focuses on the ethical design of care robots, namely how care robots can be designed and implemented in care contexts in a way that supports and promotes the fundamental values in care. By summarizing the findings from the previous chapters I arrive at a conceptual framework for evaluating care robots (or other artefacts for that matter) and practices from the perspective of good care. I refer to this framework as the Care-Centred (CC) framework given the focal role the care perspective plays in both its creation and methods of use.

This framework is then used in two separate methodologies to accomplish divergent goals: 1) to evaluate the design of current care robots; and 2) to steer the design of future care robots. Each methodology differs in multiple respects. The former is used to evaluate both a current care robot and a current care practice. In this respect, the technology has already been made and the evaluation occurs downstream in the design process. The recommendations resulting from such an analysis may be used for the improvement of future designs of the care robot in question as well as other robots that may be made. What's more, the evaluation also allows for a critical analysis of current care practices. From this angle it is possible to ask whether or not a robot will in fact maintain the same standard of care, will minimize the current standard, or will enhance the current standard.

Alternatively, I refer to the latter methodology as the "Care-Centred Value-Sensitive Design" (CCVSD) approach. It is a prospective methodology beginning further upstream in the design process, at the moment of idea generation. There is no artefact made and the methodology is intended to shape the future design, design process and implementation of a care robot. Although each methodology is presented as being separate from the other the two are inter-related in that the findings from the retrospective evaluations inform and substantiate the CCVSD approach.

The CC framework and the methodologies for using the framework, in either a retrospective or prospective manner, both pay tribute to the central thesis in care ethics, namely that the care perspective provides an orientation from which one can begin to theorize as opposed to a pre-packaged ethical theory. The framework articulates the components which require attention for analysis from a care perspective while the methodology indicates how these components are to be dealt with. The framework consists of five components: context, practice, actors involved, type of robot, and manifestation of moral elements. Each of these components will be described in detail for understanding their place within the framework from the care ethics stance.

It is true that robots used for different practices, by different users in different contexts, will vary greatly and thus so too will the ethical evaluation of such robots; however, the goal of this work is to show there are certain components fundamental to the provision of good care that must be addressed in the ethical evaluation of every care robot in order to do justice to the promotion of good care.

Table 5.1 Care-centred framework for the ethical evaluation of care robots

Context – hospital (and ward) vs. nursing home vs. home setting …
Practice – lifting vs. bathing vs. feeding vs. delivery of food and/or sheets, playing games …
Actors involved – human (e.g. nurse, patient, cleaning staff, other personnel) and nonhuman (e.g. care room, mechanical bed, curtain around bed, wheelchair, mechanical left, robot …)
Type of robot – assistive vs. enabling vs. replacement
Manifestation of moral elements – Attentiveness, responsibility, competence, responsiveness

The Care-Centred Framework

It is necessary to recognize that care must be understood in its totality, as a practice integrated into a holistic process rather than an unlinked series of actions or tasks to be fulfilled. This means that the framework and methodologies must point towards an understanding of care in this sense. In order for this approach to design to succeed, this understanding of care must be related to, and integrated into, issues of design.

Context as a Component

Firstly, one must identify the context within which the care practice is taking place. On the one hand, the context determines the structure of care, the resources available, and the various routines in place for patients and personnel. In this sense, structural context refers to the specific hospital and ward vs.

a nursing home vs. a home setting. On the other hand, context can refer to a cultural climate that plays a role in how things are seen and done. Recent research points to a relationship between religious beliefs and one's acceptance of using robots in care-taking roles (Metzler, 2008). Metzler and Lewis are investigating the hypothesis that when one believes in "a god" they may not be as inclined to accept human-robot interaction with life-like robots at an intimate level. This is because of the belief that humans cannot play God in the sense that they should not be creating things which closely resemble humans. Thus, the design of a robot for a Catholic hospital ought to take this kind of research (i.e. cultural context) into consideration for the appearance of the robot.

Similarly, the structural context in terms of one hospital ward or another is also of great importance when designing the robot. Research done by Bilge Mutlu of the University of Wisconsin, Madison shows how the same robot (the TUG® robot) used in one hospital was accepted differently depending on the ward. Workers in the post-natal ward loved the robot, while workers in the oncology ward found the robot to be rude, socially inappropriate and annoying. The same workers even kicked the robot when they reached maximum frustration (Barras, 2009). The routines and structure of the hospital ward was said to have an impact on users' responses in the sense that the neonatal ward was not as stressful or high intensity when compared to the neo-natal ward (Barras, 2009).

Insights into the structural context and the routines and resources within, also allows the design team to consider design restrictions based on said context. For example, movement corridors in homes are normally more restricted than in a hospital setting requiring more agile and perhaps smaller robots. Movement in hospital corridors does not pose the same restrictions.

Specifying the structural context in terms of a nursing home vs. a home setting is also of importance given that the prioritization of values differs. The practice of lifting in the nursing home, for example, places efficiency as a high priority (and even more so in the hospital) while in a home setting timely completion may not be as necessary without the same time constraints. In addition, certain practices, like bathing, in a home setting may not require the same demand for privacy as the hospital or nursing home setting given the lack of other patients around.

What's more, both structural and cultural contexts play an integral role when we consider the need for establishing and/or maintaining the relationship/bond between care-giver and care-receiver and furthermore what this bond is like. In a home setting, the relationship has already been formed between care-giver and care-receiver (they are often family members or spouses who already have a deep understanding of the preferences and routines of each other), thus a robot may not pose the same ethical concerns. Alternatively, in a hospital setting where daily practices are intended to establish and/or strengthen the bond between care-giver and care-receiver as well as to learn about preferences and styles, each practice serves a pivotal role in this process.

Practice as a Component of the Framework

A care practice is defined here as an identifiable moment in which the actions and interactions between and among actors (human and non-human) result in the manifestation of values. The carrying out of the practice is also how we come to understand the distribution of roles and responsibilities and thus the practice is one of the central foci in the evaluation of a care robot's impact. Examples of practices include but are not limited to: lifting, bathing, feeding, fetching items, delivery of medications/food/x-rays/sheets to the room or to the nurse, personal communication, social interaction, games and activities like singing songs or painting.

Each care practice results in the manifestation of care values; however, they mean very different things depending on the type of practice. For example, the value of privacy when discussing the practice of bathing refers to maintaining the privacy of the corporeal body of the individual – not allowing external parties to view the patient's body. Privacy for the practice of personal communication when thinking of a vulnerable patient sharing their feelings and fears amounts to keeping this information secret or the patient anonymous to respect their identity. Thus, privacy in both instances refers to a respect for the dignity of the care-receiver, through the non-disclosure of information, bodily or verbal. The detailed description of each practice is further dependent on the context within which care occurs as well as the actors or demographic involved.

Actors as a Component

Authors Borenstein and Pearson (2011) point to the differences in needs between demographics – that certain demographics will have needs that differ from others. Take robots in child care and robots in elderly care as an example. One may suggest that the element of "play" is crucial in the design of a robot in the first instance while natural language communication is more important in the design of a care robot in the second instance (Borenstein, 2011). In the same vein, authors Sharkey and Sharkey (2011) also point to differences between these two demographics only this time they stress that significance of human contact in the cognitive development of children and the significance of human contact in the reduction of stress, increase in global cognitive functioning, and decrease in risk of developing dementia in elderly populations (Sharkey and Sharkey, 2011). Hence, human contact is important in both instances, but for very different reasons. With this in mind, the design of a care robot must take the intended actors, or users, and their distinct needs into consideration.

From the care orientation, the actors involved are of great significance for structuring moral deliberation. One of the most important findings to come from the care ethics perspective is the ontological status of humans as relational. Its significance for this work lies in recognizing that the care practice which a robot will enter involves a network of human (and non-human) actors in relationship. The robot then has the potential to shift the roles and responsibilities distributed within these relationships as has been stressed already. What's more the robot will engage in relationships with any number of actors in the network.

The actors involved in a care practice will differ between a hospital, nursing home, hospice, children's hospital, or home context. In any setting, the patient may be completely dependent on others for certain practices. In a hospital setting, the actors will be the patient and any number of healthcare personnel. If the patient is receiving care in their home perhaps the actors involved are family members or a visiting nurse who is not present on a daily basis. In a home or hospital setting a patient may fulfil certain practices on their own prior to a robot assisting. This does not mean the care-receiver is entirely on their own, in the atomistic sense, but rather that the robot may be delegated a certain portion of the role of the care-receiver (as is the case with a feeding robot like Secom's MySpoon).

In each instance care-giver and/or care-receiver enters into a relationship with the robot. By entering into a relationship with the robot I do not equate human-robot relationships with human-human social relationships where empathetic and trusting bonds are formed. Rather, I am referring to a superficial/simple interaction between robot and human, which over a long period of time, and with the added dimension of expressing values, can be referred to as a relationship more so than an interaction. It is true that certain robots may have social capabilities and thus convey feelings of empathy but I hesitate to refer to such interactions as forming a relationship. I defer to the work of Turkle (2011) here and claim that the one-sidedness of the relationship makes it impossible to equate it to that of a human-human one.

It is important to remember too that the human actors are not acting alone to manifest values. They work together with each other but also with technologies already in use in the healthcare system. In nursing and technology studies, technologies have often been considered extensions of the nurse's body or self (Sandelowski, 1997). Nurses become so skilled at using the technology they do so without being distracted by the technology's presence. Of course, we must recognize there will always be a time in which the nurse learns to use the technology and appropriates it into his/her daily routine; the technology being an extension of the nurse does not happen automatically. What's more is that the nurse's role is one that incorporates the use of technologies in a variety of ways from the mechanical bed to heart monitoring devices. Thus, technologies are not only extensions of the nurse but they also mediate the relationship between the nurse and the patient shifting both the role and the responsibility of the patient and nurse in order to include the technology in the equation.

We are not speaking of interactions or relationships that occur without the use of technologies. Therefore, the question is not what happens when a care robot enters the nurse-patient relationship that is devoid of any technologies. Rather, we are speaking of a context within which technologies are already employed to a high level and the question is how will a care robot alter the existing relationships; what relationships will it form and what relationships might it interfere with?

Type of Robot as a Component

As we saw in the last chapters, defining and classifying robots is quite complicated given the range of capabilities. Some consider a type of robot according to the domain for which it is used; industrial vs. rehabilitation vs. military vs. search and rescue robots (Veruggio, 2006). For others, types of robots may be in terms of industrial robots vs. service robots. Service robotics are then further distinguished according to personal service robots (e.g. pet robot or automated wheelchair) and professional service robots (e.g. delivery robot in hospitals) (International Federation of Robotics).[1] A distinction between industrial and service robot has to do with the amount of human interaction the robot will have and the predictability or structuring of the environment that the robot is working.

For the purposes of the framework discussed here, I classify 'type of robot' with how the robot will be used among the human actors – how a role and responsibility is delegated to a robot. For example, an enabling robot is one which enables a human to perform an action previously not possible without the robot or, the robot enhances the human's performance during a task – the robot and human are working together toward a goal but the human is in control of both him/herself as well as the robot. Thus, the responsibility for accomplishing that role is a shared effort with the robot perceived in an instrumentalist way, as a tool. Robots of this type are: tele-presence robots like the RP7, surgical robots like Intuitive Surgical's daVinci®, or exoskeletons like the Hybrid Assistive Limb (HAL).

Next, a replacement robot: this robot is one that fulfils a practice in place of the human. The role of the human and the associated responsibilities are delegated fully to the robot. An example of this type of robot is the RI-MAN autonomous robot for lifting, Secom's MySpoon automatic feeding robot

1 See http://www.ifr.org/service-robots/ (last accessed 17 September 2014).

(when programmed to fulfil all tasks of feeding), or the Sanyo electric bathtub robot that automatically washes and rinses.

Finally, an assistive robot: this is a robot that aids a human in performing an action by providing a portion of the practice without the direct input of a human operator and is thus delegated a partial role and a partial responsibility. This robot differs from an enabling robot in that it does not require consistent input from a human but rather can execute a practice once given its command. Examples of this kind of robot are: the TUG® or HelpMate™ robots used for deliveries in hospitals; the Mitsubishi Wakamaru robot for monitoring and delivering messages; or the Nao robot used for medication reminders (Anderson and Anderson, 2007). In the case of the delivery robots, the role and responsibility of the delivery is shared between the robot and the human deliverer/receiver; however, the robot fulfils many steps without any input from a human user (e.g. navigating through hallways and corridors).

In addition to the type of robot identified, it is important to list the robot's capabilities as they too play a role in the ethical analysis. Will the robot be collecting information? Will this be stored? If so, for how long, how is it encoded and who will have access to it? Will the robot be linked to a telecommunications network during its functioning and thus be subject to hackers? Will the robot be engaging in physical interaction with a human (as in the case of lifting or bathing robots)? Will the robot be speaking and if so at what volume will be acceptable in a given hospital ward (or nursing home)? Will the robot autonomously return to its battery station and if so what happens when it cannot make it there on time, what de-fault mechanisms will be in place if the robot breaks down in the middle of a hallway? Many capabilities of the robot will demand specific ethical treatment to comply with the ethical standards of the healthcare tradition.

Alongside robot type and robot capabilities, the appearance of the robot is also a key concern. Arras and Cerqui report on their Swiss survey that only 19 per cent (n = 2000) of the participants preferred a human-like appearance (2005). Dautenhahn et al. (2005) report that although human-like communication is desirable for a robot companion (in their study a personal service robot was used), human-like behaviour and appearance are less important overall. This component of the framework links with the contextual issues but also raises the issue of comfort level when interacting with the robot.

In terms of the appearance of the robot, expectations of users is also central; "people expect a robot to look and act appropriately for different tasks" (Goetz, 2002), one might think it odd to have a creature-like robot helping to feed patients.

With these thoughts in mind, the appearance of the robot should be taken into consideration as a design issue in the discussion of the robot's capabilities and features as it too will impact both the acceptability of the robot as well as the role of the robot as perceived by the human users.

Manifestation of Care Values

This component of the framework refers to how the care values are manifest within a defined care practice; in a given context, with the specified actors involved. This proves quite difficult when we are aware of the many values which are expressed at any given time during any given care practice. Before I can go further I must select the fundamental care values.

Many care ethicists make clear the range of values and principles that provide a normative account for care (Vanlaere, 2011; Little, 1998; Ruddick, 1995; Noddings, 1984); however, they fall short of providing a systematic way to visualize and evaluate these principles and values. The vision presented by care ethicist Joan Tronto allows for a perception of care as a practice with stages (1993), that provides the most enticing conceptualization for engineers to work with (van Wynsberghe, 2013a). Not only that but it also provides us with a tool for labelling the fundamental care values.

Essentially, care is a cluster of values that come into being through the actions and interactions of actors in a care context (see Chapter 2 for a list of care values). Creating a standardized framework to guide the promotion of these values which applies to any care context, care practice, care-receiver or care-giver is quite problematic given the range and variety of care values discussed. Moreover, their ranking and prioritization is dependent on the context (e.g. one hospital domain or another vs. a nursing home) and practice (e.g. lifting vs. bathing). What's more, to claim that human dignity, compassion or respect for power are values to be embedded in a care robot offers nothing for the designer in terms of the robot's capabilities.

To overcome this obstacle I turn to the care ethics literature which tells us that alongside values needs play a central role in the provision of good care. The needs of the patient mark the starting point of the care process and the process then revolves around a care-giver taking steps to meet these needs. But we can't forget that the care-giver has needs too: needs in terms of resources, skills, and responsiveness from the care-receiver to understand when needs have been met as well as their own personal needs. So which needs should we be talking about?

I propose that it is possible to outline a set of needs for *every care practice*. To that end, I suggest using the criteria from the care ethics literature explored in chapter two to uncover the specific needs in any care context which ought to be met in order to ensure good care. More specifically, the moral elements given by Tronto: attentiveness, responsibility, competence and responsiveness.[2] These moral elements may be considered fundamental needs of any care practice which must be fulfilled in order to meet the requirements of good care.

2 See Table 5.2 for the moral elements and their definition.

With this in mind I am also suggesting a specific way to envision the relationship between needs and values: "values are objectives that one seeks to attain to satisfy a need" (Super, 1968, pp. 189–90). This means that, the value is the goal one strives towards and in so doing, intentionally meets a need. In other words, we begin with needs, and the values represent the abstract ideals that, when manifest, account for the needs of individuals.

Table 5.2 The moral elements and their definitions according to Joan Tronto

Attentiveness	capability of recognizing the dynamic needs of a patient
Responsibility	closely aligned with trust; requires an understanding of what one is doing and why; capability of identifying response to needs and delegation to meet them
Competence	capability of executing a means/action to fulfil the identified needs in a skilled manner
Responsiveness	capability to engage with the care-receiver regarding the meeting of their needs

With this suggestion, there are two assumptions being made: 1) every care practice will always have the moral elements as needs, independent of the care-giver and care-receiver; and 2) all care values are subsumed within the moral elements.

With regards to the first assumption, I do not claim that the moral elements are interpreted and prioritized in the same manner for every care practice but rather that they are necessary and sufficient conditions for good care in every care practice. Using the practice of bathing as an example, I suggest that this practice will always require attentiveness, responsibility and competence on the part of the care-giver and will always require a reciprocal interaction between care-receiver and care-giver for determining whether or not the needs have been met. This is so no matter what the context is.

I cannot make the same claim for the valued action of touch. Touch is important for the practice of bathing; however, I cannot say that touch is a valued action that ought to be present in every care practice. For the delivery of sheets or food to the room, there is no reason to assume that touch is required. Or, for social interaction in which a care-giver sits with the care-receiver to talk, there is no grounds on which we can say that touch is required. I can, however, make the claim that attentiveness is required for these practices and refers to a recognition of the patient's preferences for touch (or a lack thereof). Consequently, the moral elements are needs, independent of the care-giver and the care-receiver, and are necessary for good care to be achieved.

The moral elements are, however, dependent on the context and practice for their interpretation and prioritization. If we compare the practice of lifting with the practice of feeding we can see how the element of competence is uniquely interpreted in each practice: skilfully bearing the weight of another without dropping or causing pain vs. skilfully coordinating timing with placement of food and utensils. In terms of context, the practice of lifting in the hospital requires greater efficiency than the practice of lifting in a home setting where time may not be as much of an issue. Thus, although the moral elements must always be present, context and practice still play a crucial role in their interpretation, prioritization and manifestation. This recognition again reinforces the care ethics stance.

For the second assumption – that all the values are subsumed within the moral elements – I claim that every care value, from touch to human dignity, is brought into existence through the moral elements. Oftentimes, the values are analogous to a phase or moral element while other times they are expressed through the manner in which an action takes place. The value of patient safety is fulfilled through the competent completion of a practice (the phase being care giving and the moral element being competence). The valued action of touch requires attentiveness on the part of the care-giver for determining when and to what degree touch is considered necessary. The manner in which care practices take place is tailored to the specific likes of one care-receiver or another and again requires attentiveness to those preferences and competence in meeting them. What's more, paying attention to those unique preferences is the vehicle for establishing trust and allowing for successful reciprocal interaction.

In short, the moral elements may be referred to as "the fundamental values in care". They represent the values needed for each and every care practice. What's more, they include the variety and diversity of care values discussed in chapter two. As such, ensuring the elements are present or strengthened through the design and introduction of a care robot, ultimately results in a manifestation of the core care values. The differences in prioritization and manifestation of elements between practices and/or contexts are something that the (robot) ethicist must draw the attention of the designer to. Before this can happen, the designer must first be aware of the necessary elements and their manner of manifestation.

A Note on User Preferences

One point of clarification that should be made explicit at this time has to do with the normative criteria for evaluating the care practice and the care robot. Many design studies (e.g. user-centred design, participatory design, and certain versions of scenario-based design) address the preferences of users to mark both the starting point for design as well as the evaluative criteria for the design of an artefact. It should be clear from the description of the values of the CC framework that in fact the preferences of patients are not represented as part of the evaluative criteria. This is so for a variety of reasons.

First, a distinction in preferences and needs must be made clear; needs represent the essentials for functioning. In care terms, needs are defined as the essentials for the provision of good care. In contrast, preferences assume there is a choice between one alternative and another. The use of patient preferences to steer the design of the robot does not recognize the pivotal role played by the needs of the institution and the needs of care providers.

It is true that healthcare institutions revolve around meeting the needs of patients, (i.e. the needs of patients determine the actions and attitudes taken in order to meet those needs); however, in order to meet the needs of patients, the needs of the institution and the needs of care providers must come first. Care-givers must be afforded the opportunity to form a therapeutic relationship in order to meet the needs of patients. Care institutions must allocate resources efficiently in order to meet the needs of all. By designing care robots in a way that facilitates the meeting of needs of care-givers as well as the needs of the institution, the care robot is ultimately meeting the needs of patients. This is not to undermine the preferences of patients; rather, it is to insist that care providers be assisted in their role as care provider as a way to meet patient needs.

Second, the impossibility at this time of creating one robot that is capable of adjusting its behaviour based on individual patient preferences, for every patient that it encounters. Of course in the coming decades this is subject to change, at which point my claim may be to question whether or not a robot should be designed to accommodate user preferences (i.e. should this not remain the responsibility and skill of the human care givers?). Yet, the work of this book is an explicit avoidance of speculative ethics and as such I do not go into much detail in speculating possible future robots that may or may not come into fruition. For now, adjusting one's behaviour based on individual patient preferences remains a capability exclusive to human care-givers. This point also ties in with the assumptions and meanings associated with the delegation of certain roles and responsibilities to care robots.

Applying the Framework

Referring to Dubberly's discussion of design process models, he distinguished the model from the actual method of using the model – the design process. I too distinguish here between the framework and the method for using the framework. The framework points to the components demanding ethical attention and applying the framework allows me to analyse the components of good care practices with and without the presence of a care robot. Thus, I take the stance that the initial investigation of a care practice must be criticized with respect to the moral elements rather than taken as the gold standard. This is also done to avoid a kind of built-in naturalistic fallacy. In other words, how care *is* currently practiced becomes the normative standard for how care *ought* to be practised and thus built-in to the care robot.

The methodologies for use act as a "user manual" and differ in multiple respects: the point in the design process that the evaluation takes place; the kinds of recommendations that result from the differing evaluations; and, the intended audience of those recommendations (designers vs. policy makers). When used retrospectively, the evaluation is integrated into the design process further downstream and will yield recommendations for an improved design of the care robot (according to the demands of the care ethics tradition) and/or for guidelines pertaining to its implementation. When used prospectively (i.e. the CCVSD approach) the framework is intended to shape the entire design process of the care robot from the moment of idea generation to the resulting artefact and its implementation.

Retrospective Evaluations

For retrospective evaluations, the (robot) ethicist identifies and describes in great detail: the context, practice, actors, and the manifestation of moral elements, related to the care practice. This is done without any discussion of the care robot and is necessary in order to understand who is acting when, for what reason, and how values are manifest through such actions. The manner in which the practice is described pays tribute to the idea that the practice is situated within the overall process of care and that the human and nonhuman actors act together in a network to produce values.

I have already shown how care values are the result of actions and interactions and thus the component "manifestation of moral elements" aims to make explicit how the moral elements come to fruition in a network of actors (human and non-human) in a given context for a specific practice.

Of equal importance in this description is making explicit the roles and responsibilities of all actors (human and non-human) in the network (care practice) as well as how the practice relates to other practices and to the overall process of care. This is done to ensure a fluid description of the care practice as well as making explicit the roles and responsibilities that often go unnoticed. For Tronto, "any account of institutional care that fails to name explicitly the 'care-attentives' and the 'care-responsibles' allows those people, and their roles in caring, to pass unnoticed. Such not-naming contributes to the process of 'naturalizing' care relations and to blaming the care-givers who may have inadequate resources" (Tronto, 2010, p. 165). By outlining the distribution of roles and responsibilities prior to the inclusion of the robot the design team is able to understand any shifts that may occur following the introduction of the robot. Added to this itis important for the design team to understand the chain of responsibility within the overall process (who is responsible for what action/ role) as well as how and when the therapeutic relationship is formed.

Following this meticulous illustration of the care practice, the (robot) ethicist then discusses the type of robot (assistive vs. enabling vs. replacement), the capabilities, features and appearance of the robot, and the manner in which the proposed care robot is assumed to fulfil its role. The (robot) ethicist describes the practice again in great detail, only this time with the introduction of the care robot. The ethicist must indicate what role and responsibility the robot has been delegated as well as how the moral elements are manifest throughout the care practice with the inclusion of the robot. With this, the (robot) ethicist can make clear the impact of the robot's capabilities on the framework of values (i.e. the manifestation of moral elements). Relating this to the care stance and its implications, the (robot) ethicist now has the potential to uncover if, when, and how a robot interferes with the provision of good care understood in terms of the manifestation of moral elements.

Evaluating the robot against a human's capabilities is done in the case of a replacement robot, but in the case of an enhancement or assistive robot the robot is evaluated according to the role it has been delegated. This evaluation then addresses the robot's impact on enhancing or assisting the human care-givers capabilities for meeting the requirements of the moral elements.

Of particular interest is the impact on the overall process of care, the shifts in role and responsibility distribution, the variety of assumptions embedded in the design and the prioritization of needs/values that the robot assumes. Thus, the ethical assessment begins with the manifestation of moral elements and incorporates additional reflective tools to uncover a range of alternative ethical considerations.

Science and Technology scholar Madeline Akrich discusses the embedding of assumptions made about user preferences and competencies (1992). Placed in context, each care robot will take on a distinctive meaning and the meaning of the robot has to do with the assumptions embedded within. Making potential assumptions about care practices, care personnel and/or care contexts explicit helps to identify additional ethical issues to be addressed.

This description is quite useful for my reflection but an important distinction must be reiterated here, one that pertains to the difference between assumptions and the concept of values and norms. Assumptions are more about the real word, they are descriptive in a sense while values are more about what the real world ought to be like, they are normative in a sense. When an assumption is made about a value to be embedded, it does not have to be a description about what is, but could also be a claim about what values ought to be expressed, how they ought to be expressed, or what priority they ought to be given. In others words, when the built-in assumption pertains to a value, or when a valuation is being made, the result is a normative claim about what the values should be, what should be valued, or what the ideal is.

For Akrich, "many of the choices made by designers can be seen as decisions about what should be delegated to a machine and what should be left to the initiative of human actors" (Akrich, 1992, p. 216). By making choices about what should and should not be delegated to certain actors (human or non-human), engineers may change the distribution of responsibilities in a network according to a preconceived ideal.

Consequently, each robot will reflect divergent assumptions pertaining to the understanding of a care practice, the aim of the care practice and the prioritization of values manifest through a care practice. It is for this reason that I insist on evaluations of care robots on a design-by-design or practice-by-practice basis. From such an analysis the goal is to provide normative suggestions for either the implementation of the robot or an improvement in design. As such, the retrospective evaluation addresses certain prospective suggestions based on the analysis but differs from the prospective evaluation proposed here.

Prospective Design

Using the framework prospectively involves the same steps as described above but begins further upstream in the design process during which time there is no robot prototype to discuss. Instead, the discussion begins with care practices. The ethical analysis of the relationship between robot capabilities and its impact on the manifestation of moral elements, roles and responsibilities is repeatedly debated throughout the design process right up to the implementation of the robot. In this way ethics (i.e. ethical value analysis) is included throughout the entire process. I refer to this process as the Care-Centred Value-Sensitive Design (CCVSD) approach.

The first step in the design of a new product is the phase of idea generation during which time engineers, designers and roboticsts begin to speculate on potential applications of the robot. For the CCVSD approach, during this phase, the ethicist visits the hospital or nursing home to understand care in context; the number and variety of practices and their link with one another. At this time the ethicist must also observe and take note of how values are translated and ranked in context.[3] It is important for the ethicist to understand care in a specified context and to be able to explain to the engineers the meaning associated with certain care practices and the relationship one practice has with another as well as with the overall process of care. Above all, the ethicist will understand where in fact there is a relevant need for a robot's assistance.

The next step, according to the CCVSD approach, is to select a practice for which a care robot may be designed. For this, the ethicist must elaborately describe the care practice in meticulous detail to uncover: how values are manifest through the actions and interactions of actors (human and non-human); how a particular practice is related to other practices and to the overall care of a patient; areas in which a robot may provide the possibility to re-introduce certain care values; and, elements that ought to remain intact if not strengthened through the use of the robot. With this information, the ethicist collaborates with the design team to brainstorm the care robot in terms of its capabilities, features, appearance and functioning. With a prototype in mind, the practice is described once again in the same terms as above, only this time with the addition of the care robot as an actor in the practice. Thus, the ethical acceptability of the robot's capabilities (linked with their assigned roles) is studied on a case-by-case or design-by-design basis rather than making sweeping generalizations as to what any care robot in general (or as a whole) ought to be capable of.

I suggest here a specific role for the ethicist as a member of the design team to assist in the moral deliberation among designers in a way that brings about practical solutions for the design of the robot. To do this successfully, the ethicist must have experience with robots, aka. a robot ethicist, (either in his/her educational background or work experience) and must also have visited the context within which the robot will be implemented. Only by visiting the context of use can the ethicist be of assistance when describing the care practice and the resulting manifestation of values. Accordingly, only by having a baseline knowledge of robots and their capabilities can the ethicist make suggestions that are within the realm of the designers' capabilities. Most importantly, it is the ethics training of the ethicist that endows him/her with the capabilities to identify moral issues and to highlight contextually significant moral values.

Additionally, the CCVSD approach incorporates the aspect of implementing the care robot. For this, recommendations resulting from the analysis of care practices alongside the development of the robot are incorporated into the introduction of the robot in the context of use. In other words, if the robot is designed to be used in a specific way, by a specific user, at a certain time,

3 This suggestion conforms to the suggestions of Le Dantec et al. (2009) building on the traditional VSD methodology who claim that values must be understood in context (both cultural and structural) for proper interpretation.

such recommendations are expressed to the actual users in context as they become acquainted with the technology. This is to avoid the manifestation of an unexpected or random morality within the context of use.

While this may seem paternalistic at first glance, the entire aim of the CCVSD approach is to shape the care robot according to the values of the institution, an in-depth understanding of the practice and context of use, and a commitment to the care ethics tradition. Thus, the care robot is designed with a specific use in mind in order to maintain the values of the institution and a high quality of care such that allowing for random or unintended uses threatens the integrity of the CCVSD approach. This aspect targets one of the criticisms of traditional VSD concerning the disconnect between intended and unintended values which we will discuss later in Chapter 8.

Conclusion

Given the lack of regulatory frameworks in the design process of robots used outside the factory coupled with the need for ethical consideration throughout the design process of care robots, I have created a tool to guide design according to certain fundamental concepts and principles in care. This tool is meant for use in the design and development of any care robot. The focus on care practices and the therapeutic relationship comes from the context, practice and relationship within which care robots will be placed. Thus, the framework is tailored to the context of a care institution, the complexity of care practices and the central focus of the therapeutic relationship in the provision of care. Beginning with the care orientation and using the blue-print of VSD I arrive at a conceptual framework for evaluating artefacts and practices from the perspective of good care. The framework is heuristic in that each component is subject to description and interpretation (from the care ethics perspective) alone as well as in relation to the other components. Additionally, it is prescriptive in terms of directing one's attention to the necessary components and the means for their analysis.

The methodologies for using the framework retrospectively and prospectively have been described in detail in this chapter but let us continue now to see the CCVSD approach in action!

Chapter 6
Care Robots and the Practice of Lifting

Introduction

At this point we know that good care is comprised of a variety of care practices. Many of these can be labelled "activities of daily living" (ADLs). ADLs is a terms used in healthcare referring to practices like personal hygiene and grooming, dressing and undressing, self-feeding, self-transfer such as getting into and out of bed, bowel and bladder management. They are practices considered essential to one's daily functioning and are therefore routinely used to test the functional status of a patient (Krapp, 2002). ADLs serve a variety of purposes beyond the primary end of the ADL. A practice like bathing can also be a process in which a nurse helps a patient enact out their identity as a citizen (Pols, 2004). Care practices are moments during which the therapeutic relationship between patient and care-giver is either established or strengthened. Care practices meet social needs as well as physical needs (Mol, 2010; Tronto, 1993; Pols, 2004).

Given that the majority of current care robot initiatives aim at developing a robot to assist with ADLs (e.g. Secom's MySpoon assistive robot for feeding, the RIBA robot for lifting, and the Sanyo electric bathtub for bathing (Sharkey, 2012), the ethical evaluation of a care robot used in an ADL must take into consideration the multiple ends which an ADL serves in the therapeutic context. Evaluating the robot using the CC framework does just this.

The aim of this chapter is to show the relationship between a robot's capabilities and the resulting promotion of care values; a difference in one capability or mode of control is shown to dramatically shift the resulting care practice. Added to this, the chapter aims to make clear the kinds of assumptions that find their way into the architecture of a care robot.

The following chapter uses the CC framework for the retrospective evaluation of current care robots used in the practice of lifting. This is done according to the "user manual" described in Chapter 5. To recap, the practice is described in detail with careful attention to how and when the moral elements are manifest. The distribution of roles and responsibilities are also clearly articulated, along with the relationship between this care practice and others (the overall care of the patient). Following this, the robots are described and once again the practice of lifting is presented in the same manner as before, only this time with the inclusion of the robot.

To be clear, I am not describing the current practices and concluding from this their normative force, rather I am describing the current practice to articulate where and how the moral elements are manifest according to their interpretation from the care ethics perspective along with where and how certain values may have taken precedence at the expense of others. It is important to be critical of current care practices in order to uncover ways in which a robot might help promote an ideal vision of care.

The Practice of Lifting Using Human Actors

In the home, hospital and/or nursing home setting, the practice of lifting may involve only human actors for the actual act of lifting. The description of the practice is as follows:[1] the human actors are the patient and the nurse (or nurses if more than one is needed, or porters if nurses are not available) and the non-human actors are the chair or mechanical bed which the patient is getting out of, the curtain (if the patient is in bed) and the hospital room. The nurse approaches the patient and asks if they are ready to get up. With agreement from the patient, the nurse places his/her arms at the patient's waist and waits for the patient to place their arms around his/her neck or on his/her shoulders. The nurse makes eye contact with the patient at all times to cue the patient. Together, they work to lift the patient from the chair and onto the bed or into another chair (from wheelchair to dinner chair for example).

If the patient requires more assistance while confined to their bed, oftentimes porters are called on to help. In this case, the nurse and/or porters: enter the patient's room; speaks to the patient to let him/her know they will be moved; encloses the curtain around the patient's bed and; begins to lift the patient. It should be noted that there will be times when the practice doesn't happen exactly like this, when the nurse is in a hurry for example; however, the pictures presented in the two instances above are also meant to show the ideal vision of the practice.

This is one way to describe lifting when only human actors are involved for the majority of the practice (aside from the wheelchair and/or mechanical bed etc.). This is also the kind of description of a practice that leaves one with the image of a task – that lifting is a moment in the care process in which the only

---

1 The description of the practice of lifting is the result of observations in a hospital context in Ontario, Canada. This is not to say that all lifting occurs in the same manner but we may speculate that there is a certain formula to lifting that is common to most hospitals and nursing homes and it is this formula that is presented here.

goal of the task is to lift. The practice of lifting is about much more than this as we shall see once we include a discussion of the moral elements.

For this practice, we may consider attentiveness as the nurse's perception of how the patient is doing on that day, at that time. Perhaps the patient's condition is worsening and the nurse can observe this through his/her physical presence during the practice. Perhaps the patient is in a greater level of discomfort or their neurological status is worsening. Or, perhaps the patient is recovering quite quickly. Thus, not only is the nurse responsible for being attentive to the mechanistic criteria defined by the practice of lifting (lifting at a certain speed, applying the appropriate amount of pressure, etc.) but the nurse is also responsible for being attentive to the unique state of the patient on that day at that time – the overall process of care. The nurse will use this information and tailor other care practices of that patient's overall care accordingly.

The moral element of responsibility is closely aligned with the element of competence and both rely on the concepts of liability and trust. The nurse (or porters) must competently lift the patient which means lifting at an appropriate speed, using an appropriate amount of force, and lowering the patient according to the same criteria. Responsibility then refers to the capability of the nurse to be accountable and liable for accomplishing the lift with competence. The hospital too assumes a portion of liability. Responsibility in this sense demands that the nurse, porters, and/or hospital be blamed in the case that something goes wrong (e.g. a patient being dropped or hurt in the process of lifting). It is this awareness that encourages the patient to trust the hospital, nurse and porter.

The nurse is also responsible for fulfilling the practice competently as it applies to the overall provision of care: does the patient prefer the nurse to be silent or to chat? Does the patient prefer the nurse to give a certain warning before lifting takes place and to take his/her time or does the patient prefer the nurse to be as fast and as efficient as possible? By understanding what the patient prefers in this practice (i.e. having the capacity for being attentive to and understanding the personalized patient) the nurse begins to understand the unique details of one particular patient with particular needs and preferences. This is a responsibility of the nurse in terms of providing good care and is a skill (or competence) that the nurse develops over time.

It is this piece of the practice that strengthens the therapeutic bond, helping to encourage the formation of a trusting therapeutic relationship between care-receiver and care-giver(s). This relationship is necessary in the provision of good care throughout the rest of the care process. This bond is required for the patient to be honest about their symptoms, to take their medications, to comply with their care plan and in general to follow the advice of the nurse.

The trusting bond is achieved not only through attention to, and an understanding of, the unique patient and their needs. The value of trust is maintained or promoted throughout this practice in additional ways as well – through the interactions between not only the human actors but the material world as well (i.e. the practice and the meaning of the practice is a co-production of actions and interactions between actors and material objects). One might assume that enclosing the curtain acts to encourage trust, through a respect for privacy. Privacy, if we recall, refers to a non-disclosure of the corporeal dimension of an individual. One might also assume that the eye contact made between nurse and patient also acts to promote trust. Or, one might suggest that having the nurse physically present in the room encourages the manifestation of trust.

Accordingly, the nurse's presence is a causal and symbolic representation that they are responsible for the well-being of the patient and further, that this is a necessary element for trust in this practice. It is symbolic in that the nurse is a representation of the legally and culturally sanctioned policies of the hospital. It is causal in that the physical presence of the nurse demands that the nurse be liable in the case of something going wrong, according to the sanctioned policies. We might go further and suggest that a hospital must have such a policy to ensure that whoever partakes in a practice like lifting is responsible and/or liable regardless of the level of worker (i.e. a nurse versus a porter versus another kind of healthcare personnel).

This last point, the presence of the human actor for lifting seems to be the necessary criterion when it comes to the moral element of reciprocity. Reciprocity in this practice refers to the patient giving cues as to their own well-being – these may be verbal or non-verbal. The patient may make verbal recommendations about their discomfort or satisfaction and the nurse (or porters) can respond accordingly. Alternatively, if the patient cannot speak, or chooses not to, the nurse is still capable of picking up on subtle cues given by the patient like wincing or a look of fear. Reciprocity involves an interaction between nurse and patient and adjustments being made according to the attentiveness of the nurse regarding this reciprocal interaction. This has been discussed previously as "tinkering" (Mol, 2010) or "ethical sensitivity" (Weaver, 2008).

In short, the attentiveness of the nurse, as described here, facilitates an understanding of the patient as a unique person with dynamic needs and preferences. The human presence of the nurse/care-giver encourages the formation of a trusting bond. Lastly, the interactions with the material world – namely enclosing the curtain around the patient for privacy – helps to express core care values.

Nevertheless, this is not always how the practice of lifting occurs in the hospital or nursing home setting. Interestingly, although the moral elements are intact, they seem to come at the price of the nurse's physical well-being. Oftentimes the nurse does not have the physical strength for lifting multiple patients in a given day and many times there are no porters around for assistance. In these cases, to ensure efficiency within the institutional setting, nurses rely on a mechanical lift.

The Practice of Lifting Using a Mechanical Lift

The lifting of patients has proven to be quite a challenging feat for nurses. Many elderly patients in the hospital or nursing home require partial assistance for lifting themselves out of bed, or out of a chair. Alternatively, many are not capable of supporting their own weight at all and require complete assistance of a nurse to get out of bed or out of a chair. In addition, many post-operative patients require partial or full assistance for transferring themselves from the bed to a wheelchair, etc. Given that the nurse must do this for any number of patients in a shift, there is a risk to the nurse's physical safety when she/he is required to lift multiple patients in a day. What's more, many nurses are not physically strong enough to do this. As a result, nurses have opted to use mechanical lifts on the many occasions that patients need to be lifted (Li, 2004).

When using a mechanical lift, in a hospital or nursing home context,[2] the practice of lifting involves the following actors: the nurse, the patient, the mechanical lift and its remote control, the mechanical bed, the curtain to enclose the patient and the room. It should be noted that there are many mechanical lifts commercially available which will change the practice of lifting in certain respects. Nevertheless, the practice will follow a certain pattern in most instances and it is this pattern that I wish to outline.

To provide complete lifting assistance for raising the patient out of their bed,[3] a slip is placed underneath the bottom of the patient while they are lying in bed (the patient is raised to an appropriate degree using the mechanical bed). Each side of the slip is hooked onto ropes hanging from the ceiling. At the press of the remote control the ropes work to lift the patient off the bed and into the air. Using the remote, the nurse moves the patient to position them over their wheelchair and begins to lower them into the wheelchair, again using the remote.

---

2 The specific context in which I had the opportunity to observe the practice of lifting using the mechanical lift was the critical card ward of a hospital in Ontario, Canada.

3 I emphasize here that this is one kind of mechanical lift and there are many more commercially available. Additionally, this is not a statement about the performance or commitment of nurses; rather, this is a critique of the technology and its impact on the values in care.

In this picture of lifting, there is a significant risk that attentiveness of the nurse for the patient is directed more towards the mechanical lift and its remote than the human patient. When the patient is being lifted, there is no physical contact with the nurse; although the nurse is physically present there is limited chance for eye contact as the patient is raised up. What's more, in order for the mechanical lift to operate the nurse must pay attention to the remote control. When this happens eye contact and touch are not possible. In terms of trust, one might assume that the patient trusts the mechanical lift only because the nurse (whom they trust) is using it, or perhaps because of the trust they have placed in the institution they are in (I refer back to the causal and symbolic elements of trust through the presence of the nurse in the above description).

In terms of competence, we might say that the nurse and mechanical lift accomplish the lift efficiently without injury and that this satisfies the criteria for a competent lift. However, this neglects the link between competence and attentiveness – with the nurse's attention and focus directed towards the remote, or even shared with the remote, the attentiveness of the nurse towards the dynamic reactions, and/or cues, of the patient is compromised. This is not to slight the work of the nurse as I am in no way implying that the nurse should not pay attention to the mechanical lift. My intention is to point out what happens when the technology is introduced into the network.

In terms of reciprocity, many opportunities for being attentive to the responses of the patient are missed during the course of the lift when/if the nurse is watching the remote and positioning the chair. The mechanical lift is not responsible for understanding any of the unique preferences of the patient therefore it is still the responsibility of the nurse to make adjustments based on their observations of the patient. These observations, however, are minimized; she/he may not catch a sudden wince on the face of the patient indicating some kind of pain. Reciprocity between the patient and nurse, in this instance, is limited to the moment of placing the patient in their wheelchair.

From this discussion there seem to be significant risks to the ideal vision of lifting when the mechanical lift is introduced. What the mechanical lift does provide is relief for the nurse from the repetition and strain of lifting. What's more, the mechanical lift allows for efficiency in the practice of lifting.

If we evaluate the mechanical lift on efficiency alone we seem to be missing something. This is not to say that efficiency is not a value in healthcare contexts or practices. In fact, the incredible range of needs demanded of the institution requires that it be efficient in order to meet them all. Is it possible though to have a technology that provides an alternative balance to care values? Can we have a technology that will assist with the provision of efficient care while at the same time promote the wellbeing of nurses alongside the values of attentiveness, competency or reciprocity?

The current technology involved in the practice of lifting shows us how important it is for designers to understand the holistic vision of a care practice – how it acts as a moment for the promotion of care values and is linked with a patient's overall care. We are reminded that the introduction of care robots may perpetuate the trend to minimize certain care values if we do not question current practices. Perhaps more importantly though this example presents a unique opportunity to re-introduce certain values of ethical importance that may have been lost or overlooked with the introduction of the first wave of automation, in this case the mechanical lift.

Enter the Robots: Care Robots for the Practice of Lifting

There are two kinds of robots I will use to discuss the practice of lifting: 1) Autonomous lifting robots; and 2) Human–operated/exoskeleton robots. In the first group are robots like RI-MAN which now goes by the name of "Robot for Interactive Body Assistance", or RIBA. RIBA "has a teddy bear face, and can pick up and carry people from a bed to a wheelchair. It can recognize faces and voices and responds to human commands" (Sharkey, 2011).

The second group of robots, the exoskeletons, are those like HAL from Cyberdyne (Hyashi, 2005), the HULC™ (Human Universal Load Carrier) robot exoskeleton from Berkeley Robotics and Human Engineering Lab,[4] or the Honda Walking Assist Device with Bodyweight Support System.[5] This class of robots are most commonly used in rehabilitation and military contexts. For the former, the robot is intended to be used by soldiers needing to carry heavy weights/large quantities over long distances. For the latter, the robots are intended to be used by patients with severe spinal cord injuries who can no longer walk or who have limited capabilities. These kinds of robots were not originally intended to be used in the practice of lifting but recently roboticists have been exploring the possibility that it be worn by a nurse for assistance in lifting within the practice of bathing where the patient must be transferred from the bed to the bathtub (Satoh, 2009). In the same vein I propose the robot be worn by a nurse for lifting regardless of whether or not the patient is being transferred to the bath.

---

4 Accessed on 22 March 2014, http://bleex.me.berkeley.edu/research/exoskeleton/hulc/.

5 Accessed on 22 March 2014, http://corporate.honda.com/innovation/walk-assist/.

Both robots can achieve the same goal (i.e. lifting a patient); however the technical capabilities through which this is achieved differs and thus changes the way in which the care values are met along with the vision of the resulting care practice.

An Autonomous Robot for Lifting

The autonomous robot is autonomous in that it is capable of lifting a patient and carrying him/her from one place to another without being controlled by a human operator. Thus, the robot is considered a replacement robot. This robot is designed to work directly with humans and is programmed for safety considerations (e.g. such as speed) and the materials which are used for its structure are pre-tested for human contact.

In the case of RI-MAN, the robot has a humanoid appearance, meaning the robot has a head, eyes, a nose and arms. In the case of RIBA the robot is creature-like. These robots are intended to work in a hospital, a nursing home or in someone's home.

It may be possible to have lifting robots that are delegated a certain portion of the practice, and fulfil that role autonomously, but those will not be the focus of this section. For the purposes of this discussion, the interesting point is the complete delegation of the role and responsibility of lifting to the robot. It should be noted that these robots are currently, in the year 2014, in the very early stages of development. This is ideal given that the recommendations resulting from the following evaluation may be incorporated into its current design process and/or implementation.

The practice

The practice of lifting in the nursing home or hospital context using an autonomous robot involves the robot entering the patient's room and fulfilling the practice entirely on its own. The robot must recognize the person to be lifted and respond to a voice command to lift. Such a response may or may not be in the form of verbal communication (this is at least the hope for this robot in the future but at present the robot is still in the early stages of development). Requiring that the care-receiver give the command to lift assumes that this person is capable of giving a voice command that the robot can understand; in the case of frail or elderly persons this may not be the case. It may also be possible that the patient is not fluent in the language of the institution. These are design considerations related to usability.

The autonomous robot assumes that the manner in which the patient is lifted will not differ between patients but will be standardized; the patient must be lying in a position in which the robot can recognize it is the patient to be lifted and in which the robot is capable of manoeuvring. As the robot lifts the patient it is hoped that it may alter the force, speed or angle at which it lifts. These are all design considerations for current and future versions of autonomous robots for lifting that must be met in order to make the robot safe and successful for the goal of lifting.

Reflecting on the autonomous robot and the moral elements

Immediately with this description we can see how the design of the robot prioritizes the value of efficiency over other elements like attentiveness and reciprocity. The robot, being designed to lift a patient in a standardized manner, leaves no room to be attentive to the individual needs of the patient. The robot may be capable of fulfilling the action of lifting according to a mechanical description of the practice (the angle at which one is lifted, the speed with which one is lifted, the force with which one is lifted); however, even such a mechanical description of the practice is dependent on the unique patient. Not every patient is lifted in the exact same way. Without the ability to be attentive to the changing needs and/or status of the patient, the practice cannot be tailored accordingly and thus the competence of the care-giver is in question.

In terms of responsibility, the robot has been delegated the full role and responsibility of lifter. Thus, if something were to go wrong, the robot would be liable for damages. But if the robot cannot be liable according to the traditional conception of liability (i.e. a human care-giver would be sued or fired), then is it possible to say it is really responsible in this practice? If the robot cannot be responsible, or liable, then should hospitals) be allowed to delegate such a role to the robot in the first place? The difference between the mechanical lift and this robot here has to do with the presence of a human worker versus the absence of a human, respectively. Without a human present, to whom, and how, are we to assign responsibility? I leave this point for now and will take it up again in the next chapter when I discuss delegating responsibilities to robots in general.

In terms of reciprocity, the patient's placement in a chair, toilet or back on the bed is the only source of reciprocity: only if and when the patient makes it to their destination does the robot know it has fulfilled its task. This kind of reciprocity is quite different from when the human actors were able to interact and observe the responsiveness of the patient throughout the entire practice. At the moment, the robot is not sophisticated enough to acquire cues about the patient's satisfaction, or lack thereof, during lifting nor can it tailor its performance accordingly. One might wonder whether this is the kind of recommendation for designers or whether this kind of role ought to remain in the domain of a human care-giver.

The answer to the above question has to do with how the robot impacts the manifestation of moral elements and values as they relate to the overall care process (i.e. the multiple practices within care that are linked via the values).

For example, human presence was necessary for achieving attentiveness of the nurse to the personal preferences of a particular patient towards the mechanics of the practice of lifting (speed at which one lifts, angle and force) but also for gathering information about the daily changes of a patient. This information is then used by the same nurse involved in other care practices or is passed on to inform the other nurses caring for the same patient. So it's not just about whether or not the robot is able to perceive all the little cues of the patient but whether or not that helps to facilitate good care when understood as an overall process.

On top of this, the human presence of the nurse creates a moment for establishing and/or maintaining the bond of trust between nurse and patient. With this in mind it is important to ask: what could happen if the nurse is removed entirely from this practice?

Let's imagine a scenario in which a patient, Nathan, is in the hospital recovering from surgery. The first day after Nathan's surgery a robot comes to lift him from bed and take him to the bathtub. Nathan was quite hesitant at first but was reassured that the robot had been tested and would not drop him. By the third day Nathan began to trust that the robot would not drop him and by day five Nathan was fine, if not happy, with the presence of the robot for lifting. In this scenario, we can assume that Nathan may have had fears about the use of the robot to begin with but these fears were overcome through repeated, successful, use of the robot. Added to this we can assume that Nathan expects the robot to assist with lifting and has grown accustomed to the manner in which the robot does so (in a precise, standardized manner without asking any questions or engaging in any discussion). We might even go so far as to suggest that Nathan is satisfied not only with the robot being there but also *how* the robot accomplishes its goal.

Aside from Nathan's satisfaction with the lifting robot, what happens to Nathan's relationship with the nurses looking after him if he hasn't had as many encounters with them? Will Nathan trust the human nurse after having such encounters with the robot? Will Nathan even want to interact with the nurse if he has grown accustomed to the robot? If Nathan has not had any moments to bond with the nurse will he be honest about his recovery and symptoms with the nurse when she/he comes to check-up on him at a later time?

It is possible to suggest that the nurse will have other opportunities to bond with Nathan but will they provide the same kind of information? The information obtained through the practice of lifting, just as in the practice of bathing, provides information about the neurological, physiological and sociological status of the patient. Can the robot be programmed to gather this kind of information; maybe the robot could be programmed to ask certain questions and the answers will be transmitted back to the nurse in charge of the patient. This suggestion poses other problems for the nurse and patient:

what if the patient can't answer the robot? What if the patient doesn't trust the robot and doesn't want to answer? What if the patient lies to the robot? What if the patient is worried about who will have access to that information and consequently doesn't respond to the robot?

Let's not forget that it's not simply a question of information that is given verbally – much of the nurse's work deals with those cues and signals that are not given verbally but that the nurse's training allows him/her to pick up on (Pols, 2004). The robot is not making connections between one care practice and another for a particular patient in its head (aka internal programming) in the sense that it is not comparing Nathans fragility while being lifted in the morning with his clammy skin during bathing in the afternoon. When the nurse is delegated a role and responsibility in the multiple practices of a patient's day, they assume overall responsibility for the patient and are tacitly reminded of this as they conduct themselves. Even if the technology were sophisticated enough to make these kinds of connections we must ask whether or not it should be delegated a role for which making such connections is required. This last point again brings us back to the idea that a robot shouldn't be given capabilities that demand a moral agent to take responsibility for the consequences of the action.

One might say that there will be persons who prefer to talk to the lifting robot or who prefer to have the robot lift them in the hospital or nursing home. This point ties in with what I said in the previous chapter regarding preferences; preferences do not provide the ethical foundation for care. Preferences differ depending on culture and time, and can even change overtime for the same individual. Given the requirements for good care in care institutions, namely the formation of the therapeutic relationship and the fulfilment of the moral elements, it is of the utmost importance that the moral elements remain intact over the patient's preferences. The moment of pleasure that the patient feels sharing a conversation with the robot may have a detrimental impact on the nurse's ability to provide good care at another moment.

Of course one may respond to this and say that the good of the patient is the goal of institutional care (Pellegrino, 1985; Vanlaere, 2011) and therefore each patient should be able to share a conversation with the robot if they so desire. I do not want to deny a person a conversation with a robot but I do claim that the needs of the care practice (and ultimately the care-giver) must be placed at the fore to ensure that the system of the institution is functioning efficiently as well as in accordance with the values of the institution. With this statement it is clear that I too prioritize efficiency as a top value in the healthcare tradition, but how I interpret and arrive at an efficient system is based on the entire system; how values are manifest within that system, and the interconnections of the system.

A change of context

Let's take a look at the same robot using the same evaluative criteria only this time with a change in context. Interestingly, the results of the evaluation differ. In a home setting in which a close family member or care-giver is the one providing daily care services there may not be the same need to ensure moments for establishing and maintaining the relationship or the trusting bond; these may have already been established through the existing relationship between the care-giver and care-receiver. In this case, there is not the same pressure to ensure the presence of the human care-giver as a means for ensuring a link with the nurse and the overall care of the patient. The care-giver at home is most often already aware of the personal preferences of the care-receiver.

Making this situation even more interesting, the care-giver may be the spouse, child, or another family member of the care-receiver which makes the care-receiver feel quite vulnerable and powerless to be in such need. In these cases, perhaps the more dignified means for lifting is in fact the use of an autonomous robot. In this way, the robot is actually able to provide the care-receiver with a sense of autonomy and/or dignity. This, of course, is only when the robot is designed to meet the necessary requirements of performing the practice safely and competently.

This change of context reiterates just how important the context is for understanding the hierarchy of care values. This is the exact kind of recommendation to be taken into consideration during design but also for the policy makers involved in the implementation and guidelines governing the introduction of the new technology.

A Human-Operated Robot for the Practice of Lifting

While the previous group of robots were capable of replacing the human care-giver that would normally lift the patient, this type of robot is meant to assist the human care-giver with their task. It is an enhancing robot in that it augments the skills of the human care-giver. By reading the biometric signals of the care-giver, the robot is able to bear the burden of the weight of whatever the care-giver is lifting without *feeling* the burden of the weight. This could be a patient, a bed, a heavy box, etc. I say feeling here to indicate that the person doing the lifting doesn't experience the physical (muscle) stress or strain of lifting.

The practice of lifting using the human-operated robot proceeds in a similar manner as the practice of lifting with human actors. The nurse, wearing the robot, enters the room, indicates to the patient that it is time for lifting, encloses the curtain around the patient and begins to lift the patient with careful attention to the speed, angle and force with which lifting occurs for this particular patient. The nurse's attention is not directed towards the suit or a remote control and consequently, she/he is capable of engaging in eye contact with the patient to pick-up on any non-verbal cues. Their presence also allows them to converse with the patient if desired and/or needed. They touch the patient in a manner of speaking; certain parts of touch occur through the robot's apparatus rather than through human-to-human touch.

The care-giver decides when the care-receiver needs to be lifted, at what speed, from which angle, and with or without social interaction. For the latter, reciprocity is something that happens between the care-giver and care-receiver in real time by verbal and non-verbal cues which are picked-up by the care-giver. This means that the nurse can ask the patient how they are doing while he/she is being lifted. The nurse is present to observe non-verbal cues which supplies them the opportunity to learn about the patient.

As for responsibility and competence, these elements now become shared endeavours between the human and the robot. The robot here has the role of weight bearer and is responsible for carrying the weight of the patient but for nothing else. Thus, a certain amount of competence for the skilful completion of the practice is delegated to the robot. This situation is quite similar to the one in which the mechanical lift is used. The care-receiver and care-giver must both trust the technology – responsibility for the safety of the practice becomes a hybrid event between the human care-giver and the robot helper. Hence, the nurse must be capable of competently using the robot. The robot must accurately translate the movements of the care-giver into its own movements with synchronicity just as the surgical robot translates the movements of the surgeon's hands into movements of the robot hands. Therefore, a portion of the responsibility for lifting is delegated to the robot as is a certain level of skill.

Such robots are not endowed with tactile sensation or force feedback, at this time, and one might question whether or not this is the kind of information the robot must be capable of acquiring in order to adequately label the robot competent. Alternatively, as in the case of surgical robots, a lack of force feedback has demanded that the surgeon is trained for a new style of surgery. Thus, the surgeon conforms to the technology rather than the alternative.

Given the role and responsibility the robot has been delegated, the overall successful completion of the practice of lifting remains the responsibility of the human care-giver such that they are accountable, and liable, for a failure. So what does this mean for the overall process of care if the human care-giver and the robot share certain roles and responsibilities rather than delegating all roles and responsibilities to the robot (or to the human for that matter)? To begin with, the human care-giver is present throughout the entire care practice and is (required to be) focused on the needs of the patient while monitoring the patient's preferences.

One of the main questions of interest in this evaluation is whether or not the robot will pose the same problems as the mechanical lift – will it detract the attention of the care-giver from the patient to the robot? In response, when the robot is worn by the nurse there is no remote for the nurse to focus on. The robot essentially becomes an extension of the nurse in his/her role of lifting the patient just as a stethoscope becomes an extension of the physician in his/her role of assessing the patient. In this way, the assumption (and hope) is that the nurse's attention will be directed entirely on the patient whom she/he is lifting. The use of the robot also frees up the nurse's attention. She/he can make eye-contact with the patient and can engage in conversation because she/he is not straining to bear the weight of the patient (as may be the case when lifting without any assistance).

By ensuring the presence of the nurse for this practice, not only are the moral elements safeguarded but so too is the linkage between this practice and other practices the nurse engages in with this patient. We can safely say that the nurse is able to learn about the patient during this practice and to use that information for the patient's continual care. Seen from this angle, the robot has the potential to re-integrate lost values when compared with both the mechanical lift (i.e. the values of attentiveness and eye contact) and with traditional lifting involving a human nurse alone (i.e. relieves the physical burden of the nurse).

Conclusion

It is only through a deeper understanding of what care values are and how they are manifest throughout a care practice that we come to grasp the impact a design might have on the care practice. This chapter was meant to outline the relationship between the technical capabilities of the robot and the resulting expression of care values. To do this I chose the practice of lifting and how different robot capabilities resulted in a different evaluation from an ethical stance. Most importantly, the tool to use for the ethical evaluation is the CC framework proposed in this book.

The autonomous robot reflects a vision of the practice of lifting which does not require any of the values traditionally involved; human touch, eye contact, human presence. If these values are normatively understood and recognized as only possible through human-human interactions, then this demands that a human be present for the practice of lifting in all instances. However, as we saw here, context plays a role. It is possible to suggest that in the context of the nursing home or hospital, where "good care" depends on the relationship between the care-giver and care-receiver, human contact for a practice like lifting is required. This is in part due to the vulnerability of the care-receiver while being lifted as well as the need to form a bond between care-giver and care-receiver. Alternatively, in a home context in which a relationship between care-giver and care-receiver is already established and strengthened, the need for human presence, eye contact and touch for the practice of lifting may not be as pertinent. Moreover, when the care-giver is a spouse it may be preferable not to have the human present. Thus, both design and integration into the healthcare system are of importance here.

Alternatively, the exoskeleton or human-operated robot, reflects an understanding of a care practice as one in which the aim of the practice is not solely to lift the care-receiver from one place to another but is a moment to establish a bond and convey other care values. The vision of care presupposed in the design of the exoskeleton care robot is one in which individualized care with a human care-giver present at all times for all parts of the care practice, is the overall aim. Efficiency is still a priority; however, it is achieved through meeting the needs of a care-giver by contributing to the element of competence (enhancing the skill with which the care-giver may perform their role), attentiveness, (enabling the care-giver to perceive the minute cues of the care-receiver through the practice of lifting), and responsiveness (closely aligned with attentiveness but also embodies the reciprocal dimension of the relationship). Consequently, by demanding the human's presence for the task of lifting, the robot pays tribute to the holistic vision of care and the intertwining of needs and values.

I cannot say whether this is the epistemic aim of engineers, but can only point to the potential meaning that the robot may take on through pervasive use, and the presupposing assumptions directing such a meaning. This is not to say that RI-MAN ought to be disregarded or labelled as unethical – a variation in context changes things. Clearly, decisions concerning the use of a robot and its ethical implications are many-sided and complicated; they demand an understanding of the specific context and users for anticipating how the elements will be served to their greatest potential.

It is one thing to make an evaluation of a robot during its development or once it has been made but it is quite another thing to make suggestions for a novel robot that takes ethical considerations to heart right from the moment of idea generation. The following chapters are dedicated to proposing new robots in a way that incorporates ethical analysis and an ethicist in the design process of the robot.

Chapter 7
The Future Design of Care Robots: The Care-Centred Value-Sensitive Design Approach

Introduction

Sherry Turkle presents her book *Alone Together: Why We Expect More from Technology and Less from Each Other* (2011) to mark the opportunity we have to shape robotic technology in a way that protects cultural values that we hold dear to us and want to safeguard. According to Turkle, robots offer us the opportunity to re-think our conceptions of what a relationship is and what it means to be in relationship with someone, or something, else.

Following up on this I propose that *now* marks the opportunity to shape the design and development of care robots in a way that safeguards the values which form the very foundation of the care tradition. Moreover, that the entering care robot provides us with the opportunity to re-evaluate care at the institutional level and to design the robot in a way that reinforces the purpose of such institutions. This opportunity for re-thinking and shaping requires a bridging of gaps so to speak, a coming together of disciplines in a way that allows for an understanding on the part of engineers as to the impact this technology can (potentially) have from the ethicist's perspective, but also to encourage the translation of ethics into a tangible format for engineers to grasp. Like Turkle, I am encouraging a stewardship of values. Beyond Turkle's claim, however, I am presenting a format for accomplishing this feat.

This chapter is meant to illustrate how the prospective design of care robots can, and should, happen through an incorporation of ethics into the design process. I refer to this approach as the Care-Centred Value-Sensitive Design (CCVSD) approach. I will show: at what stage in the design process the framework is meant to be used (idea generation and onwards to implementation); how to proceed with the framework (its method for use); and finally how to interpret the reflections from using the framework (recommendations for design and/or implementation). In order to do all this I propose two novel robots; the *wee-bot* and the *roaming toilet* robots.

One of the most important goals of this chapter is to discuss the moral status of the care robot. This dimension has concrete consequences for the designers of care robots, and the resulting care robot design, and will therefore be discussed at the beginning before I can proceed further with the CCVSD approach. Lastly, as an important component of the CCVSD approach, I will also illustrate the role of the ethicist as a member of the design team of a care robot.

Designing Future Care Robots: Using the CCVSD Approach Prospectively

As laid out in Chapter 5 the CCVSD approach is a prospective methodology for designing a care robot in a manner that incorporates the core care values of attentiveness, responsibility, competence and reciprocity. There is no current robot to evaluate and therefore the initial value-based analysis results in recommendations for designers to begin the making of a care robot such that it embeds care values in the most promising way. With the initial prototype, the robot may then be re-evaluated (in context) again using the framework to arrive at additional recommendations for designers or alternatively recommendations for policy-makers.

To show the utility of using the framework for the design of future robots I proposed a novel robot called the *wee-bot.* I have used this robot in previous work so I will draw from this to describe the robot and the initiative for the robot (van Wynsberghe, 2013b; 2014).

> In the pediatric oncology ward in a hospital the nurse is responsible for a variety of activities: cleaning the patient, maintaining a sterile environment, keeping up to date with new research protocols and treatments as well as others.[1] One responsibility of the nurse in this ward is to test the urine of the children undergoing chemotherapy for the presence of chemo toxins found in the urine. Essentially, the nurse is checking to ensure that the therapeutic intervention – the chemotherapy – is present (the next step is to test if the intervention is working and whether or not higher or lower levels of the toxin are required). To do this the nurse puts on protective clothing as the chemicals are quite toxic and can cross the skin barrier, enters the bathroom, takes a sample of urine and tests it. With this information the nurse reports the findings back to the oncologist.[2] (van Wynsberghe, 2013b)

---

1 For more information see: "A Framework for developing practice in pediatric oncology nursing", http://www.rcn.org.uk/__data/assets/pdf_file/0004/115870/001062.pdf (last accessed 6 March 2015).

2 Information gathered and idea generated from the author's visits and observations in a paediatric oncology ward in a hospital in London, Canada.

The idea for the robot came from observations and interviews in a hospital setting in which it was made clear that due to time constraints nurses sacrificed their own wellbeing in order to ensure that the test still happened: "The wellbeing of the patient comes first and we don't have time to fully cover up so we just enter the bathroom covering our mouths and do the test" (personal communication).[3]

With this in mind I posed the question: can a robot be integrated into this practice as a way to enhance the safety and wellbeing of the nurse while at the same time preserving the wellbeing of the patient? The CCVSD approach was used to investigate this question by exploring a novel robot prototype and deliberating how certain robot capabilities would alter the expression of values (van Wynsberghe, 2013b).

In short, the robot is intended to enter into a patient's room and collect a urine sample from patients undergoing chemotherapy. How this happens and the capabilities the robot has to make this happen involved a great deal of discussion. Of particular interest was the relationship between the robot's capabilities and the delegated role and responsibility that came to light (van Wynsberghe, 2013a; 2013b; 2014). This has invited an explicit discussion of the relationship between moral agency, moral responsibility and the role delegated to a care robot. I pause now before proceeding with the next step of the CCVSD approach in order to address this important issue.

Moral Agency, Moral Responsibility and the Care Robot

The moral status of the care robot is something I have implicitly made reference to throughout the previous chapters but would now like to explicitly address and discuss because of its significance in the prospective design of such robots. The question of interest here and within the field of robot ethics in general is: Is it possible to consider robots as moral agents? Many scholars have provided conditions under which a robot can be considered a moral agent (Sullins, 2006).

Conclusions on the robot's moral agency and moral status depend on your concept of both of these terms. Typically speaking, moral agency is required for moral responsibility. When the robot is delegated certain roles in which it is required to make decisions according to the programming of an ethical theory it is presumed to be a candidate for moral agency. The robot is delegated moral responsibility such that it ought to be responsible/liable for the outcome of its decisions. But can technologies, specifically robots, be responsible for their actions?

3 Personal communication with nurses in paediatric oncology in London, Canada.

Here, I take a look at three prominent theories of moral agency: the organic view, the standard conception, and the morally intelligent view. According to the first two views, a robot cannot be a moral agent and therefore should not be delegated actions where moral responsibility is required. According to the third view, moral agency and moral responsibility are separated from one another; you can be a moral agent without assuming moral responsibilities. Therefore, a robot can be considered a moral agent; however, given its lack of intentions, the robot cannot be held morally responsible for the consequences of its actions. In what follows, a closer look is taken at these three possibilities.

The Organic View of Moral Agency

According to traditional conceptions of a moral agent, the criteria for agency is specified to include capacities for empathic reasoning, sentience, experiences, consciousness etc. One of the more traditional views of moral agency is the organic view. From the organic view, only a genuine organism (human or non-human animal) may be considered a candidate for intrinsic moral status and thus be considered a moral agent. This has to do with the belief that moral thinking, feeling and action arise organically out of the biological history of the human species (Torrance, 2008). From this, of course robots may have the capabilities for high level reasoning but cannot be considered full moral agents due to their inorganic make-up.

Is it the case that robots will never be granted moral consideration? Will there ever be a time when robots will be granted moral consideration in so much as the interests of keeping the robot functioning will be taken into consideration against the interests of other human actors in a network? Moral consideration refers to an entity deserving of a certain kind of moral appreciation without being capable or required to act in a certain way (infants and animals would fall under this categorization). Torrance questions whether or not moral consideration is the necessary and sufficient criterion for moral agency. If a robot deserves moral consideration can we then conclude it to be a moral agent. Further, what is the criterion or criteria to conclude that a robot is deserving of moral consideration? Is it sentience or some kind of potential as in the case with animals and infants respectively? To tackle this issue Torrance speaks of a distinction in terms of moral producers, those that are a source of moral reflection, and moral consumers, those that deserve moral attention.

In the latter category, moral consumers, fall infants and non-human animals; groups that we recognize to have a certain degree of sentience but who are not capable of moral reasoning at a high level. Those that fall within the former category are full moral agents capable of sophisticated moral reasoning along the lines of both intellectual and empathic reasoning. Intellectual reasoning has to do with a weighing of the pros and cons of all the details of the situation –

a consequentialist approach if you will. Empathic reasoning, according to the organic model of moral agency, refers to the capability and the inevitability to incorporate a kind of "affective or empathic identification with the experiential states of others, where such empathic identification is integrally available to the agent as a component in its moral deliberations" (Torrance, 2008, p. 510).

Robots cannot, at this time, exhibit this kind of empathic rationality which has been shown to be both a value in care and a pre-requisite for the good/ethical care-giver. Of course, roboticists and computer scientists may counter this claim and suggest that at a time in the future it may be possible to outline the process of thinking in order to achieve a kind of empathic reasoning and to program a care robot accordingly. However, this raises the question of why we would bestow such capabilities, and the associated roles and responsibilities that are attributed with such capabilities, in a care robot – just because we can does not necessarily mean we *should* program robots with moral reasoning capabilities.

Torrance suggests that given the organic view of moral agency, natural humans are full moral beings and artificial humanoids may be deemed as "courtesy moral beings" (Torrance, 2008, p. 520). This presents an interesting caveat: robots cannot be considered full moral agents but perhaps there is a sense of moral status that the robot will have once placed in context. Thus, we are left with the idea that care robots may be granted a kind of moral character or status based on the context within which they are placed, but what this entails is not quite clear given that robots do not have the same capabilities as full moral agents (i.e. human beings).

The Standard Conception of Moral Agency

In contrast to the organic view of moral agency, the standard conception of a moral agent refers to: "beings who are capable of acting morally and are expected by others to do so" (Brey, 2014, p. 125). Thus, "moral agents are beings that are: 1) capable of reasoning, judging and acting with reference to right and wrong; 2) expected to adhere to standards of morality for their actions; and 3) morally responsible for their actions and accountable for their consequences" (ibid.). Here, there is no indication as to the physical make-up of the agent but only to the capabilities, expectations and associated responsibilities of a moral agent.

An agent is a moral agent when the intentional states that it cultivates and the subsequent actions it performs are guided by moral considerations. This requires a capacity for moral deliberation, which is reasoning, in order to determine what the right thing to do is in a given situation. A capacity for moral deliberation requires a capacity for reasoning and knowledge of right and wrong. Moral deliberation typically results in moral judgments, which are judgments about right and wrong. It also frequently results in intentions to perform certain actions that are held to be moral, and to refrain from performing actions that are held to be immoral (ibid., p. 126).

Based on the above definition, one may be inclined to ask what is meant when one considers a moral agent to be "guided by moral considerations"? Furthermore, what could this mean for a robot? Is it a reference to the designers' intentions or the reasoning capabilities of the robot? According to this view, it is once again problematic to include robots within the category of moral agents for a number of reasons. First, robots only have intensions in so far as they have been programmed. It is the intensions of the designers and not the intensions of the robot that are considered for agency. Would this require that we consider the designer to be responsible and liable for any negative consequences of using the robot? Further, can we reasonably demand this of robot designers? Even still we could not consider the robot itself to be a moral agent according to the standard conception as the robot is not generating its own intentions and acting on those – it is acting out the intentions of the designers.

We must consider too that a robot cannot be held responsible or liable for its actions insofar as it cannot be punished for bad consequences. As for the reasoning capabilities of a robot, certain roboticists believe it is possible to program robots to have highly sophisticated reasoning capabilities making the robot intelligent enough to be considered a moral agent. This way of thinking brings us to the morally intelligent view of moral agency.

The Morally Intelligent View of Moral Agency

Dominant proponents of this view include Luciano Floridi and Jeff Sanders who claim that artificial intelligence opens new avenues when speaking of moral agents (Floridi and Sanders, 2004). Specifically, technologies with highly sophisticated mechanisms for reasoning, capable of interacting with their environment, acting in an autonomous fashion, learning and adapting to their environment, ought to alter the discussion of moral agents. Their goal is to expand the category of moral agents so that it includes sophisticated technical artefacts, rather than to alter the concept of morality such that artefacts and humans engage in a practice of hybrid morality. Within this conception, Floridi and Sanders aim to disentangle the relationship between moral agency, accountability and moral responsibility. They argue that moral accountability is a necessary but insufficient condition for moral responsibility.

According to their view, a moral agent, and ultimately a robot, may be considered a moral agent insofar as it may be considered accountable for its actions (and thus subject to censure); however, once again it may not be held responsible for its actions given that it lacks the intentions guiding it to make said decisions (ibid.).

With this view, once again we are left with the issue of moral responsibility open; even if the robot is considered a moral agent given its sophisticated abilities for reasoning, learning from and adapting to its environment, it is still only accountable for its actions and not responsible.

To be clear, there is no agreement among scholars of robot ethics regarding the status of robots as moral agents but all signs point to: no, robots are not moral agents. As such, robots cannot bear moral responsibility and need not be programmed to have ethical reasoning capabilities. Aside from this point, some robot scholars and roboticists still argue that robots can and should be endowed with sophisticated ethical reasoning capabilities given the roles and contexts in which the robots will be placed.

If we agree to this point and program the robot with ethical reasoning capabilities then, according to Floridi and Sanders, robots may be allowed into the category of moral agency. This is only possible by separating the concept of moral responsibility from moral agency. Consequently, this changes the focus of the debate. Instead of focusing on the robot as a moral agent we need to reinforce the relationship that moral responsibility shares with the provision of good care and ask whether or not it is possible to provide good care without the element of moral responsibility?

Moral Responsibility and Care

In every chapter of this book I have made reference to the responsibility of care workers in their role as care provider and the significance of this element for the provision of good care. The issue of responsibility is of the utmost importance in healthcare contexts and in the therapeutic relationship. The professionalization of medicine and nursing is grounded on this fact. According to care ethicist Joan Tronto responsibility has more than one conceptualization in the care domain. First, responsibility has to do with liability – taking the praise and/or blame for the outcome of a care action. Although technologies play a crucial role in the provision of good care they are used as additions to the care process, ultimately it is the care providers and institutes that are morally responsible for meeting the needs of care receivers.

Second, Tronto also states that the concept of responsibility in care is about much more than merely taking the praise and or blame for actions. It is about having caring intensions, *caring about* patients and their wellbeing. In short, care must be fulfilled by an agent capable of intentional states and of assuming moral responsibility; a human moral agent.

As we have seen in the above discussion regardless of whether or not one considers the robot to be a moral agent the robot cannot assume moral responsibility for its actions. But this conclusion does not pay tribute to the significant role/impact that the robot will still play/have through its inclusion in a care practice. So, how can we conceptualize this role and/or impact of the robot and furthermore what does it mean for the design of a care robot?

The Moral Artefacts View of Moral Agency

According to the moral artefacts view, a technology has a moral status given that it bears a moral impact on the actions and decisions of humans in a socio-technical network. Both humans and artefacts are referred to as actors or actants and are called as such when the program of action inscribed in their technical content enforces a moral rule – meaning they are capable of steering moral behaviour in humans and/or influencing moral outcomes. But in the traditional form of Actor-Network Theory (ANT) both the human and non-human actors have the same status. Accordingly, the care robot would have the same moral status and associated responsibilities as the human actors and as we have just seen this cannot be the case for the care robot.

If we claim that the robot still holds a moral influence on the practices of the human actors in the network but that the robot and human do not maintain the same moral status, and further that the robot is not capable of higher moral deliberation, nor should it be (based on the normative aspects of care), then what do we call the robot? For this I turn to the structural ethics approach of Philip Brey; we call the robot a moral factor. Brey nicely bridges the standard conception of a moral agent (which claims that only humans are fully capable of being moral agents given that taking responsibility for moral actions is a necessary condition) with the belief that technological artefacts do in fact have moral influence on the behaviours and actions of human agents (the moral artefacts view). In this conception, the status of a moral factor mirrors that of a "moral impact agent" according to James Moor (2006).

In short, the approach of structural ethics allows for the recognition that technological artefacts (what he calls moral factors) bear a moral character but cannot be responsible, that responsibility always falls on the human agents in the same network. For Brey, "moral factors shape or influence moral actions and outcomes" (Brey, 2014, p. 10).

> Moral factors can be positive or negative, measured against a moral rule or principle. A positive moral factor is one that contributes positively to a moral principle being upheld, whereas a negative moral factor contributes negatively. In addition, factors can be accidental or intentional. An accidental moral factor is one that happened to contribute towards a moral outcome in a particular arrangement. An intentional moral factor is one that has been intended to contribute to an outcome in a particular way. (Ibid.)

> Moral factors can also be outcome-oriented or action-oriented. In short, the difference between the two has to do with whether or not a moral factor influences, positively or negatively, the behaviour of a human agent or a moral outcome (a moral outcome being the event or state-of-affairs). (Ibid., p. 11)

Intuitively, one would conclude here that a care robot ought to be programmed according to a conception of it being an intentional positive moral factor.

But what does it mean to program a care robot according to a conception of it being a moral factor? It is possible to look to the work of Wendell Wallach and Colin Allen or James Moor to begin a discussion of moral agency as it relates specifically to robots. For Wallach and Allen a distinction is made between operational morality vs. functional morality (Wallach and Allen 2008; 2010; Wallach, 2010). The morality of a robot is measured according to two dimensions – sensitivity to values (along the x axis) and autonomy of the robot (along the y axis). These two axes are independent from one another.

Wallach and Allen expand their idea of morality for robots by referring to the classification of James Moor: ethical impact agents, implicit ethical agents, explicit ethical agents and full ethical agents. An implicit ethical agent refers to a machine whose designers have attempted to decrease the negative ethical impact of the machine in terms of safety and reliability issues. Such a class of machines are what Wallach and Allen refer to as "operationally moral" (or operational morality); the morality of the designers (values, norms, etc.) is embedded into the design of the system such that through the use of the system certain values are promoted (i.e. Value-Sensitive Design).

In contrast, explicit ethical agents are those machines that can reason as part of their internal programming. This grouping of intelligence is what Wallach and Allen refer to as functional morality and there is quite the range of machines that may fall within this broad category. In contrast to operational morality/implicit ethical agents, the "ethics" for explicit ethical agents (classified as functional morality) comes in as a capability of the machine rather than the exclusive programming of the designers. This then demands the question of when it will be possible to conclude that the machine has acted in a way that was not the intentional programming of the designer, as the field of machine learning and autonomous systems is now wrestling with (Wallach and Allen, 2008; 2010; Wallach, 2010; Moor 1995; 2006).

But such a discussion does not address the problem of moral responsibility, it merely provides a way to classify or label the robot according to its ability, or lack thereof, for ethical decision making. This is not the discussion I want to engage in, rather, I want to insist that the robot be considered to have an ethical or moral impact given its influence on the decision making of humans. Given that the robot can never bear moral responsibility it should never be delegated a role for which moral responsibility is attributed. In short, give the robot operational morality *only*. Design the robot to be a positive moral factor *only*. This conclusion places restrictions on the roles the care robot can be delegated and as such the capabilities the robot has must be carefully decided upon according to the role they presume.

Back to the Future! Deliberating Robot Capabilities

Thank you for indulging me on my lengthy side-step to discuss the moral status of the care robot! This was necessary as it plays a direct role in the kinds of capabilities the designer ought to give the care robot. This brings us to the next phase of the CCVSD approach; deliberating robot capabilities. At this stage, the goal is to discuss different robot prototypes with varying robot capabilities and envision what such robots would look like in practice (according to how they would impact the manifestation of moral elements and the distribution of roles and responsibilities). Let us now return to the discussion of the *wee-bot* robot and see what this means in practice.

What Does This Mean for the wee-bot Robot?

If you remember, I began by proposing the robot and briefly indicating why it would be used and what it is needed to do: the robot is intended to restore safety of the nurse by collecting a urine sample of patients. To do this it must enter the patient's room and collect a urine sample. This sample must be stored in a sterile environment and either tested or delivered to the lab for testing. When we translate this into design considerations we may consider three robot scenarios;[4] each scenario presents three separate robotic platforms with varying capabilities:

1. "A mobile robot that can travel through the hospital corridors and elevators avoiding obstacles via its sensors (thus, autonomously) but that is human-operated for urine retrieval and testing" (van Wynsberghe, 2014);
2. "A mobile robot that not only travels throughout the hospital autonomously but also travels inside the patient's room and collects the urine sample autonomously" (van Wynsberghe, 2014);

4 It is possible that there may be many more options that I haven't thought about and I would warmly welcome those. This discussion is not intended to be exhaustive but to illustrate for the reader what the approach consists of and to propose possible prototypes.

3. A mobile robot that acts autonomously for travel and sample collection but can only do its task (of sample collection, storage, testing or delivery to lab) when commanded by the nurse.

Each of these three prototypes presents a robot with different capabilities which results in a different vision of how the care practice will proceed. What's more, each assumes a different stance in terms of the robot's moral status as moral agent – it is only in the second scenario that the robot is delegated the entire role for the practice. As such, in this scenario it is the robot that calculates when to enter the patient's room etc.

Balancing Robot Capabilities with the Moral Elements

The CCVSD approach ties in the care values alongside the question of responsibility and with each robot scenario comes a different impact on the promotion of the moral elements. When discussing (1), the robot's role is to physically collect the urine sample with the guidance of the human operator. The robot is not responsible for determining: what time the collection ought to take place, which patient the sample comes from, or any other specific details about the collection. These factors are all decided by the human nurse in charge of controlling the robot. The robot is essentially an instrument or tool to assist the nurse.

In this picture the overall responsibility for the safe and successful completion of the task lies with the human care giver. Not only does the human still monitor the entire practice but he/she is still actively engaged in the actions of the practice with the robot as an assistant. He/she would be able to indicate if technical difficulties had occurred but would still be held liable for any problems that were to take place. In the analysis of this robot, one might point towards a potential problem in that the robot could minimize the efficiency of the care practice and detract the attentiveness of the nurse from the patient to the robot.

The second scenario (2) can be suggested to mitigate these concerns. In this scenario the robot takes over the entire role of urine collection and testing and thus the associated responsibilities. As such, the robot is delegated the same amount and kind of responsibility as was originally delegated to the human care giver. The robot is responsible for: determining when to do urine retrieval and testing; informing and interacting with the patient whose urine is being collected and tested; collecting and testing the urine sample; and, passing on the test results to the appropriate oncologist.

Given this picture it is possible to assume that the nurse may not feel needed for this practice and may not be present when the robot is acting. What would happen if the test is not taken at the right time or the results are not sent to the appropriate oncologist? Would the nurse even know that something went wrong? In the instance that something goes wrong and we can place accountability on the robot who is to blame and thus liable?

With these concerns in mind the final scenario is presented (3), to balance the distribution of responsibilities within the nurse-robot network; the robot is endowed with autonomous capabilities for travelling throughout the hospital as well as for sample collection and urine testing. The difference lies in the fact that it is necessary for the nurse to identify themselves to the robot before the robot can begin their task.

> Identification could be through voice commands, facial recognition, retinal scans, finger print analysis etc. This gives the robot permission to enter the room but also ensures that the nurse is present and responsible for the practice. To strengthen this interaction, the robot could be programmed with semantic links endowing the robot with the capacity of knowing 'why' it must ensure the presence of the nurse. The robot may also be designed such that when it leaves the hospital room it must also interact with the nurse prior to sending the information to the oncologist. What's more, once the information has been sent to the oncologist the robot requires that the nurse 'sign-off', in a manner of speaking, before the robot is able to leave the room. (van Wynsberghe, 2013b, p. 438)

With this design recommendation comes a significant shift in the amount and kind of responsibility delegated to the robot: the robot does not make any autonomous decisions regarding urine testing but is there as an assistant to carry out the actions of the nurse. The responsibility of the nurse is still to ensure that: the collection of urine is taken, the test is made, and the results are passed onto the appropriate oncologist. The robot is meant to act as an extension of the nurse, to carry out the decisions made on behalf of the nurse without putting the nurse in harm's way. Making reference to the types of robots I use for the framework we could consider this to be an enhancement robot.

Summing Up: The Phases of the CCVSD Approach

The example of the *wee-bot* robot just discussed was meant to illustrate the kind of didactic process that the CCVSD approach encourages. We can identify certain phases that are not necessary mutually exclusive but iterative: 1) Proposing the idea of a care robot to justify why, where and when it's needed; and 2) Deliberating robot capabilities by envisioning the robot in practice. For the second phase robot capabilities must be balanced with their impact on the promotion of care values (i.e. the moral elements) along with the distribution of roles and responsibilities.

Most notably, this example was meant to show how the addition of one or two robot capabilities for recognition of the nurse completely altered the landscape of responsibilities in the robot-nurse network. The final robot scenario presented was able to balance the range of values needed to promote good care: the wellbeing and safety of the nurse along with the moral elements.

Following this of course the robot prototype must be built and tested with users before implementation. Before we engage in a discussion of implementation let us look at another example of a novel robot.

A Care Robot for Waste Removal: The *Roaming Toilet*

I now switch to an entirely new robot to, once again, show the utility of the CCVSD approach. I stress that this robot, just like the *wee-bot* robot, does not exist; rather, this is a novel idea for a robot using current robot prototypes and capabilities.

I present this robot, and the discussion around the robot, from the ethicist's perspective as a member of the design team. During the phase of idea generation, the ethicist may have the role of visiting the care context to understand the needs of care workers and care institutions and to present an idea based on this. The resulting discussions then revolve around such needs and whether or not a robot can meet these needs without sacrificing the care values of the institute. Throughout the deliberation of the care robot's capabilities, the role of the ethicist is to bridge the gap between the technical designers and the needs of the institute. With this in mind the ethicist must paint a picture of how the care robot will fit into the practice. Most importantly, the ethicist must be critical of the proposed robot and point out: conflicting values, value trade-offs and potential mis-uses of the robot.

The Practice

In a hospital and/or nursing home context, patients who are not able to get themselves to the bathroom have a bed pan and will excrete into the pan. At certain points in the day this pan is then changed. What also happens with patients is a soiling of the sheets/linens which then have to be changed. This happens periodically throughout the day. When a nurse is aware of this, they will change the sheets, and/or empty the bed pan, and will carry the sheets to the cleaning station where laundry is done (or a drop off point on their particular ward). Alternatively, the nurse will empty the contents of the bed pan in the bathroom or will again bring the bed pan to a cleaning station.

To do so, the nurse will enter the room and enclose the curtain around the patient to ensure their privacy. The nurse will speak to the patient and inform him/her what she is about to do. If removing the bedpan, the nurse will do so by adjusting the covers and shifting the patient (if necessary) remove the bedpan and empty the contents into the toilet of the bathroom. If removing the linens the patient will need to be removed from the bed (in many instances) and placed in a wheelchair while the nurse replaces the sheets. The soiled sheets are then taken to the laundry room or a drop-off point on that ward.

In these contexts, safety, sanitary conditions and/or sterilization are of great importance. Maintaining safety may be thought of in terms of how the nurse removes the bed pan and/or sheets and the care with which they handle the patient. Safety is also related to the sanitary conditions of the context: it is not safe to keep the patient in an unsanitized environment. Carrying the sheets/bed pan through the halls can be unappealing for the nurse but more importantly may pose a sanitary risk in the institution – others may be exposed to the bacteria in the faecal matter. Thus, the manner in which the current practice occurs may pose a risk to the high standards of sterilization required in the context. This practice appears to be one in which a care robot could provide a welcome assistant to the unappealing job of waste removal and at the same time could increase the sterility of the hospital or nursing home context.

The actors in this practice are: the patient, the care provider, the bedpan (if one is being used), the soiled sheets (if there is no bedpan), the mechanical bed, the hospital room, the hospital corridor for transporting materials to a garbage or linen room, the garbage or linen room itself, and the service staff working in either of these rooms. All of the actors work together to manifest the moral elements.

When describing this practice according to the values, i.e. the moral elements, we must recognize this practice as having a role in the overall care of the patient. There is much information to be gained through the observation of excretion (indicative of someone taking their medication, indicative to someone's eating patterns, indicative to someone's recovery post-surgery). Accordingly, attentiveness is thought of in terms of the nurse's ability to observe the excretion patterns of the patient. The nurse must be attentive to the colour, size, smell, etc. of the faeces, the differences between days and times in a day and so on.

Additionally, although the end result of waste removal here is to literally remove the waste from the patient's immediate surroundings, it is also a moment in which the relationship between nurse and care-giver is fortified. I refer to strength in the relationship because this is quite an intimate practice between nurse and patient; it is a moment in which the asymmetry of the relationship is quite visible as is the patient's dependence on the nurse. Attention to the asymmetry in power is a responsibility of the nurse. The nurse must be attentive to the vulnerability of the patient and must ensure the dignity of the patient through his/her behaviour and response to both the excretion and the removal of excretion from the room.

One might also suggest that in getting to know the patient through this particular practice the nurse is also aware of how to treat the patient in other practices. This practice is linked with the overall process of care for the patient given that the nurse is made aware of how fragile and/or vulnerable the patient is and may take this into consideration for other care practices, like bathing or lifting. If the nurse observes a high level of discomfort in the patient, the nurse may choose to be extra sensitive when the time for bathing comes.

Consequently, attentiveness in this practice refers to the nurse's capacity to observe the excretion patterns of the patient but also to incorporate the patient's existential state, in terms of his/her vulnerability, in the treatment and removal of the excretion as well as in the treatment of the patient during other care practices.

Competence refers to the nurse's ability to accomplish the practice in a safe manner while achieving the goals of attentiveness as described above. It is possible to suggest that when the nurse carries soiled linens through the halls, other healthcare staff and visitors may potentially be exposed to bacteria found in faecal matter, urine or vomit.

Responsibility is related to attentiveness and competence in that the nurse is the care-giver who is delegated the role for notating the excretion patterns of the patient. This presupposes that the nurse is present to observe the excretion of the patient. Responsibility in this practice is thus expressed through the presence of the nurse, not necessarily during the moment in which it occurs but during the moments following the act.

Reciprocity is not solely about verbal communication between nurse and patient but also incorporates a corporeal communication in a manner of speaking – the patient communicates through the information provided from their body; their faeces, urine and/or vomit. In addition to this, however, the nurse and patient may also communicate verbally throughout the practice as a way of fostering the relationship. This will be decided by the nurse depending on the comfort level of the patient – there will be patients that would prefer not to speak at this time, patients who aren't capable of verbal communication and other patients who would prefer to engage in conversation.

This practice can happen multiple times in a day and is an integral component of the care of a patient in terms of the physiological information gained by the nurse but also in terms of a strengthening in the relationship between nurse and patient. When we take a closer look it is possible to suggest that some values may conflict: ensuring the presence of the nurse for the wellbeing of the patient may conflict with the dignity of the nurse. The moment the nurse carries the soiled linens through the hospital corridors is a moment in which their comfort or dignity is lessened. One might even suggest that this uncomfortable moment also translates to the patient: the patient feels awkward and uncomfortable having to be cared for in this way but may also feel the discomfort of the nurse. It cannot be easy to care for patients in this way and not show signs of discomfort. In fact, the potential that healthcare providers (more so in the case of family members of patients) resent having to care for patients is one of the motivators for care robots in the first place.

Additionally, one might also wonder whether or not there is a more sanitized alternative to the carrying of soiled linens throughout the hallways. This value conflict may or may not have been intentional, but rather a product of an efficient system where an alternative has not presented itself. With this in mind, the ethicist asks whether there can be an option that will still allow for the expression of moral elements but may relieve a certain portion of the discomfort experienced by nurses and patients in this practice.

Enter the Roaming Toilet *Robot*

With the goal of creating a positive moral factor for this practice, I propose a waste removal robot. I distinguish this robot from current robot prototypes, namely the Dustbot© (Networked and Cooperating Robots for Urban Hygiene) developed at the Scuola Superiore Sant'Anna's CRIM Laboratory in Pisa, Italy. The Dustbot© is intended as a garbage removal robot for outdoor areas which can operate at all times of day.[5] It has varying degrees of autonomy based on the function for which is will be used and whether or not it is being used in an area populated by human which is needs to avoid or interact with.

The waste removal robot that I am proposing here, that I call the *roaming toilet*, may use the same robotic platform as the Dustbot© or may use a platform like Aethon's TUG® robot. Regardless, it should be capable of manoeuvring through hallways and elevators autonomously (i.e. to travel to the destination at which it is needed). This merely means that the robot has preloaded information about the environment (e.g. maps) to allow it to navigate without direct input from a human controller. The robot is equipped with a sterile compartment into which the excretions from the bed pan (urine, faeces and vomit) may be placed. These are kept in a sterile compartment to ensure that any particles do not escape.

Based on the lessons learned from the *wee-bot* with respect to the delegation of responsibility, we can already make certain claims about the capabilities of the *roaming toilet*. If you recall from the above section there were three robot prototypes with varying capabilities. In the end the third scenario allowed for the manifestation of all moral elements while maintaining a delegation of responsibility to the human care giver: A mobile robot that acts autonomously

5 http://www.dustbot.org/index.php?menu=home (last accessed 30 March 2014).

for travel and sample collection but can only do its task (of sample collection, storage, testing or delivery to lab) when commanded by the nurse. I suggest the same idea for the *roaming toilet*.[6]

The robot will autonomously travel throughout the hospital, navigating corridors and elevators avoiding obstacles and people, but should require human input at certain points in order to fulfil certain roles. This is to ensure that responsibility lies not with the robot but with a human care giver. The robot may be programmed for routine visits to hospital rooms but is also capable of responding to an urgent situation in which it is needed. In the case of the latter, the nurse will call upon the robot via a computer command.

In either case, the robot will arrive at the patient's room and will receive a command from the nurse to open its compartment and allow the nurse to load the soiled bedpan or linens. A clean bedpan, stored in another compartment on the same robot, is then given to the nurse to replace the dirty one. Once the nurse has done this, he/she will give the robot the command to leave the room. The robot then travels through the hospital corridors to bring the soiled linens and excrement to the cleaning station. At the cleaning station the robot's contents are emptied, bed pans are cleaned and the robot is loaded with a new collection of sterile bed pans. This may be done by the nursing staff or the cleaning staff already present in the hospital. To avoid the risk of stigmatization or undignified feelings of patients I suggest that the robot be used for all linen removal so that hospital staff and visitors do not attribute the use of the robot with an unpleasant event but rather with normal operating protocol.

The *roaming toilet* robot has many advantages: it is available for pick-up 24 hours a day, 7 days a week; the robot is never tired from a double shift; and it will keep record of patient information which can be passed from one nurse to another (and/or physician). The robot's greatest advantage is the safe, sanitized and efficient transport of linens and bedpans throughout the hospital corridors.

What's more, to increase the dissemination and fluidity of information, the nurse may be required to verbally or manually input information about the patient during the time the bedpan/waste is removed. The robot will keep track of the time of day and number of times it has visited a patient's room in a day/week/month. This information is stored to allow the other nurses and physicians access. Thus, the robot acts as a tool for linking healthcare professionals about the status of the patient and any potential changes on a day-to-day basis (i.e. enhancing the overall attentiveness of the care personnel).

This type of design recommendation naturally introduces questions of privacy; how long will such information be stored in the robot? Who will have

---

6 I agree that it is rather convenient that I may suggest the same or similar platform for both robots but this is certainly not meant to indicate that all future care robots will have this same platform or these same capabilities.

access to that information, and so on? These privacy concerns are no different from the concerns related to the sharing of medical records in general and I would recommend that such records are not available to anyone outside the care team of the patient (i.e. the care-givers listed as being responsible for the patient).

Additionally, in the deliberation of capabilities, or the balancing of capabilities with the promotion of care values, I would recommend that the information be stored while the patient remains in the hospital. After the patient is discharged from the hospital the information may be kept in a database but the patient's identity remains anonymous (encoded through a patient number and not a name or perhaps the latest in cryptography would allow one to assign a pseudo-identity[7]). If the patient re-enters the hospital, their information may be retrieved. If the patient is deceased, the information may be (or, must be) destroyed. If a healthcare professional wishes to use the information for future studies such permission must be obtained by the patient prior to its use.

The Roaming Toilet *and the Expression of the Moral Elements*

In the above section I presented the reason for the *roaming toilet* in addition to a suggestion for the capabilities or the robot. I have already begun the deliberation of the robot's capabilities but let us focus now on how the capabilities impact the moral elements.

By ensuring that the robot requires the nurse to input the materials into the robot's sterile compartment, the robot ultimately ensures that the nurse still bears the responsibility for the practice. Additionally, the nurse is still able to observe certain patterns of excretion of the individual patients and as such maintain attentiveness. What's more, the insurance of the nurse's presence also contributes to the element of reciprocity. If the robot were designed to remove excretions or soiled linens on its own without the nurse present, the nurse would not have the opportunity for observation, reciprocity or attentiveness.

In terms of the manifestation of competence in this practice, the nurse is still afforded the opportunity for fostering his/her own competence, or attentiveness, of the individual patient, but the robot offers an opportunity to increase the level of competence when understood in terms of safety. By carrying patient excrement in a sterile compartment, the robot decreases the risk of transmitting potentially air-borne bacteria throughout the hospital. Thus, when compared with the current practice, the robot re-prioritizes the values of safety defined by a reduction in exposure to potentially dangerous bacteria.

---

7 For more information on cryptography and its use for protecting identities you may look to the ABC4Trust project: https://abc4trust.eu/.

Although maintaining the nurse's presence plays an integral role in fostering attentiveness, competence, responsibility, reciprocity, and the continuity of care, it also plays a role in maintaining the relationship between nurse and patient. Ensuring the nurse is present at various moments in the day allows for communication between the two throughout the practice, whether this communication is verbal or a physical pat on the arm. Most notably, the nurse is relieved of his/her role to carry the materials throughout hospital hallways and elevators and it is possible to suggest that the delegation of this role to the robot increases the comfort level and dignity of the nurse.

Added to this, because the nurse will not need to leave the room immediately to remove the soiled items (the robot will instead do this), this nurse may have more time and may feel more inclined to sit and spend this time with the patient. As such, the incorporation of the robot neither disengages the nurse from the overall care of the patient nor does it remove an opportunity for the nurse to learn more about the patient (in terms of their preferences and their state of well-being be it physiological or emotional).

The importance of following instructions

Let us imagine for a minute a scenario in which the nurse does not reach the room quickly enough and perhaps there is a family member present when the robot arrives for a routine pick-up. If the family member has been present for the patient's care on a daily basis, the patient and family may wish to allow the family member to load the tray, especially if this is more comfortable for them or if the nurse is not present to immediately do so.

The above scenario may seem harmless enough, especially if the patient would feel more comfortable having their family member load the bedpan into the robot. In truth, family members already empty some bedpans when visiting their family members. But having anyone other than the nurse load the tray could break the chain of care. If the nurse is no longer responsible for this practice how is he/she to build the relationship or be attentive to the patient's physiological status? What's more, who would be to blame if anything went wrong with the bedpan collection or with the robot? A family member could be held accountable for a problem but could not be held liable.

This potential for misuse of the robot enforces the necessity of a design recommendation to ensure the role and responsibility of the nurse (or another care-giver, for example a porter). In this instance, one may want to equip the robot with the feature of facial recognition or voice recognition. The robot is then capable of detecting the presence of particular persons/care-givers. Consequently, the robot maintains the chain of responsibility and acts to reinforce the role and the associated responsibility of the nurse. As in the case of the *wee-bot*, this robot too may be programmed with semantic links to endow the robot with the capacity for knowing "why" it must ensure the presence of the nurse. Thus, a redundancy is built-in to the system (i.e. an added mechanism for ensuring the presence of the nurse).

Alternatively, maybe the robot ought to allow family members to empty and/or replace bedpans. If this is the case, and it would be up to the institution to decide this, then the robot may be programmed to recognize a family member as being permitted to empty or replace bedpans. Added to this, the robot may also be required to store information about when it was a nurse emptying or a family member emptying the bed pan.

Conclusion

This chapter was meant to sketch the CCVSD approach as a tool in the design process of potential future care robots. In order to do justice to the overall intentions of the approach it was necessary to discuss the concepts of moral agency and moral responsibility as they relate to a robot. The discussion of moral agency and moral responsibility here highlighted the impossibility of claiming that a robot is a moral agent capable of assuming moral responsibility for the outcome of an action. This does not mean that the robot has no ethical impact. Quite the opposite, the robot carries a significant ethical force by its impact on the actions and decisions of the socio-technical network of the care practice, care team and care institute.

The fact that a care robot cannot assume moral responsibility poses limitations on the kinds of roles it may be delegated. By understanding how a shift in one capability changes the amount of responsibility assigned to the robot it becomes possible for robot engineers and designers to limit the amount of responsibility delegated to the care robot through a careful selection of its capabilities. This was demonstrated in the use of the CCVSD approach for the *wee-bot* robot and the *roaming toilet* robot.

The CCVDS approach is made up of phases for the ethicist and designer. It begins with proposing an idea for a care robot. This phase requires understanding a care practice in context and being able to identity and explain the practice according to the moral elements. Once this is decided the next phase is to suggest robot capabilities to meet the goal of the care robot. This requires envisioning the robot in practice and how it will impact the expression of moral elements. With this in mind the ethicist must encourage a balancing between capabilities and the promotion of care values.

Up to this point we now have a clear idea of how values can be embedded into a care robots and how they come to be realized in practice – through the use of the robot. Of particular interest in the example of the *roaming toilet* was the potential for misuse of the robot if the intended use was not followed. This potential brings our attention to a recent criticism directed at the embedding of values in general: a disconnect between those values that are intended and those that are realized in context. As we have seen through the CCVSD approach, the care robot is assumed to have an intended use in an intended context. It is under these conditions that the values come to be manifest. But, it may be the case that these conditions do not present themselves. What is to be done in such a case? This will be the focus of the next chapter when I discuss the implementation of the care robot.

Chapter 8
Conclusion: Implementing Care Robots with Care

Introduction

After the initial phases of the CCVSD approach during which the idea is presented and deliberated, the robot prototype is built and tested for use and usability. Once this is complete the robot is ready to be sent off to its buyers for use. This chapter is meant to discuss this final stage as a component of the CCVSD approach.

One of the greatest challenges when working on the embedded values concept lies in realizing these values in the actual use context of the artefact (van de Poel, 2014, p. 19). This is quite a daunting remark considering that this book is dedicated to the embedding of values; one would hope that when such lengths are taken to ensure that the values of ethical importance to a care practice are being specified into design requirements they will in fact be realized in practice. Because of this I want to address this issue now and do so through a discussion of the implementation of the care robot.

Although there exists a wide variety of studies concerning the domestication of technologies (how technologies take on meanings through their pervasive use), there is little work done on the ethical implementation of technologies – the bridging of design studies with domestication studies. My goal is to show that given the normative force of the CCVSD approach and the assumptions related to how the robot ought to be used, these insights should be translated into both the policies governing the use of the robot as well as how the robot is first introduced into its context and network of users.

What Happened to the Values?

To refresh your memory, the foundation of this work rests on the embedded values concept – the idea that there are values embedded into a technology (in this case a care robot) such that through its use those values will be realized. The values at the centre of the discussion in this book have been the moral elements of attentiveness, responsibility, competence and reciprocity. Consequently, the CCVSD approach aims at embedding those values into the resulting care robot. But, van de Poel and Kroes (2014) point out a very real criticism facing such an approach; values often intended by engineers are not realized once the technology is used in practice.

Van de Poel and Kroes (ibid.) agree that values may in fact be intended and embodied in a technical artefact but that does not guarantee that the value will be realized in context if the technical artefact is not used as it had been intended. It is possible that when an artefact is used in an unintended manner, or in a novel context, that it does not realize the intended values. It is also possible that when the artefact is used in a different context it may lead to the realization of different values.

With this in mind it becomes clear that to ensure a technology promotes the values of the healthcare tradition there are two general steps: 1) you must ensure that the technology embeds certain values (the focus of the last seven chapters); and 2) you must ensure that the robot is used in its anticipated way. For the second point I point towards the practice of implementation.

Implementing Care Robots with Care

This book presupposes that when technologies are introduced into a use context they are thought to express values through interactions with human users and other nonhuman (f)actors in the socio-technical network. Domestication is a phenomenon in Science and Technology Studies to describe the enactment or performative character of artefacts (Akrich, 1992; Latour, 1992; Silverstone, 1992; Sorensen, 2006; Jelsma, 2006). The phenomenon of technology domestication explores how an artefact blends in with existing norms and meanings but also how the artefact co-produces new norms and meanings.[1]

Both the phenomenon and the study of domestication refer to something that is in the process of happening and the observance thereof, or to something that has happened and the denoting thereof. No normative approach to domestication exists to date. For authors like Latour it is not possible to predict the impact of a technology until the technology is embedded and generates a sense of meaning. In contrast to this I suggest that we don't want the care robot to enter its context of use and create a novel, unanticipated or unintended morality. Instead, we want the robot to be used in the manner in which it was intentionally designed, a manner grounded in ethical consideration and foresight. If the CCVSD approach embeds a morality then the implementation of the robot can help to ensure the morality is realized.

---

1 Explaining and studying domestication rests on the semiotic approach of Akrich and Latour – script theory. Actor-Network theory comes into play here in that the visibility of the script, and the meaning of the artefact, is observed through the lens of the network, its actors and their interactions.

Implementation is the carrying out, execution, or practice of a plan, a method, or any design for doing something. As such, implementation is the action that must follow any preliminary thinking in order for something to actually happen. In an information technology context, implementation encompasses all the processes involved in getting new software or hardware operating properly in its environment, including installation, running, testing, and making necessary changes. Implementation in a healthcare institution also refers to the policies and guidelines structured according to a technology; how it ought to be used and what constitutes misuse. More often than not, implementation sciences in a healthcare context refers to the incorporation of evidence-based medicine into medical practice. In short, the idea behind the implementation of the care robot is that it is introduced into its context of use in a manner that ensures it is used as it was intended to be.

The care ethics perspective is integral when outlining a care practice and the components of the care practice but also provides insight when it comes to the implementation of the robot – the care ethics tradition fosters a dialogical approach, one in which roles and responsibilities are made explicit through dialogue among actors. What's more, for Tronto the purpose of the care institution is to be a guiding force for the multiple care practices within (2011). From this perspective, it is not only important to incorporate the components of a care practice into the design and development of the robot but the components are integral at the moment of implementation too.

Making reference to the implementation of care robots then, implementation refers to: the introduction of the robot in its context of use, to its intended users, for its intended practice, in its intended manner. Thus, this is the stage during which actions are taken to ensure that the intended values embedded into the robot are realized once the robot is used in context. Guidelines and policies outlining the use of the robot are also structured around the deliberated, criticized and agreed upon intended values. In other words, guidelines and policies are put into place to ensure the care robot is used in the intended manner.

So what does such an implementation look like? I suggest the following scenario:

> The ethicist enters the hospital context prior to the introduction of the robot. The staff members are brought together to discuss their current roles with respect to the care practices of said context/ward, and it is explained to them how the robot's design is a result of certain intended values (namely attentiveness, responsibility, competence and reciprocity). The interpretations of these values and their prioritization, according to the design team, are also outlined for discussion. The actors partaking in the discussions are those that will interact directly with the robot (e.g. the nurses using the wee-bot in pediatric

> oncology or the nurses who will use the roaming toilet) as well as indirect users of the robot (e.g. the doctors who will receive information from the robot as to the concentration of chemotherapy drugs in the urine of the patient or the support staff who will empty the contents of the roaming toilet).
>
> In these meetings, the ethicist uses the components of the framework to structure a discussion among participants: the ethicist describes his/her (and the design teams) interpretation of how the practice proceeded prior to the robot in terms of how and when values are manifest along with the distribution of roles and responsibilities of those involved. The ethicist then goes on to discuss what the robot is capable of doing, what role the robot will take on (according to the vision of the design team), what responsibility the robot is being delegated and how values might shift. The ethicist encourages a discussion among care workers as to the robot's role once used pervasively in the context.
>
> This forum facilitates an opportunity for discussion of the potential unintended uses, the misuses of the robot, and/or the anticipated tensions in values and how they might be overcome. By engaging the healthcare workers in a dialogue about the robot, their roles and responsibilities are made explicit as well as where and how the robot fits in to their conception of roles and responsibilities. Once this has been accomplished, the robot is ready to enter the context and practice. This is done with the assistance of technical support indicating how to use the robot, what it is capable of and what its limitations are.[2] The ethicist meets once again with staff after a short period of time to discuss the new pattern of the practice and to work out any unresolved intuitions or conflicts with the robot.

One might be asking at this point "what an interesting turn from the design of the robot to a focus on communication among a team of people who may or may not come into contact with the robot". This is for a number of reasons. The first reason has to do with a recognition once again of the care process – the vision of care being a holistic process, each action linked with the others (Tronto, 1993; Tronto, 2011). This is in contrast to the flat view of care actions as tasks – separate individual tasks serving only one purpose. When recognizing the holistic vision of care one is reminded of the interconnectedness of actions. Thus, if the robot performs one action, other actions and actors will be impacted. By discussing and preparing for

2 There is an opportunity at this point to engage in assessments about the interactions of the care robot and the users, both direct and indirect. These assessments may involve the preferences of users, objective criteria or more subjective criteria like the meanings associated with the care robot once used in context.

this, the hope is to anticipate and troubleshoot potential problems but also to make explicit the shifts in roles and responsibilities to avoid confusion or liability risks. This of course can only be done when conventional roles and responsibilities have been made explicit.

The second reason has to do with how roles and responsibilities are distributed among human and nonhuman actors, in particular the limits of delegating roles and responsibilities to care robots. I claim that essentially "care" remains a human domain: a human activity to be fulfilled by a human, enhanced through the use of a care robot. One of the main fears expressed by care workers concerns the risk to their roles and to their overall profession. It should be made clear to all involved with the design and use of care robots that the domain of care is a human domain and care robots are there to support the human staff members. Such an initiative can be achieved when the care robot is designed and implemented according to the CCVSD approach.

Tracking Intended Values

That main argument of this chapter goes as follows: given that the robot's design carries with it an assumed morality, one which has been reflected on and decided upon throughout its design process, the introduction of the robot ought to be done in a way that ensures the manifestation of the presumed morality (the distribution of roles and responsibilities, manifestation of care values, etc.). Thus, the implementation of the care robot is a carrying out of the CCVSD approach to ensure that the intended values of designers, along with the intended use of the robot, are made explicit to care providers.

An additional insight provided by van de Poel and Kroes (2014) relates to the tracking of embodied values. This is where the implementation of the care robot and the domestication of the robot meet. The authors claim that "the advantage of defining embodied values as the value that is realized if an artefact is properly used is that embodied values become more directly traceable and that engineers are better able to verify whether their designs embody the intended values" (van de Poel and Kroes, 2014, p. 19). Engaging the ethicist in the context of use creates both the opportunity to steer the use of the robot (according to its intended use) and the opportunity to observe the morality of the care robot once integrated into practice (i.e. the domestication of the care robot). The latter point allows for a tracking of the intended values. In short, the ethicist is able to compare the intended values with those values that are expressed following the introduction of the care robot.

Implications for Policy

From the scenario presented in this chapter, in addition to the aforementioned points of consideration, one may also take suggestions for hospital and/or nursing home policies concerning the standard implementation of care robots, namely, that: 1) the care robot must always be introduced in a dialogical manner after bringing together the direct and indirect users of the robot; 2) an ethicist involved in the design and/or the development of the robot (with an understanding of the technical details as well as the intended values) must guide the implementation; and 3) the ethicist ought to track the domestication of the robot as a means for tracking the intended values of the care robot.

These recommendations are in addition to current hospital practices in which a kind of technical support accompanies the introduction of a technology into a hospital domain. The hope is that the arguments made in this book point towards the necessity for ethical support in addition to technical support given the moral context within which the robot will be placed.

Conclusions of the Book

There are many conclusions and benefits to be drawn from the work in this book. Firstly, this book presents a concrete methodology for the creation of future care robots, the Care-Centred Value-Sensitive Design (CCVSD) Approach. This approach explicitly integrates the care ethics perspective in the design process of care robots by: structuring the design process, shaping the resulting design, and steering the implementation of future care robots. Given the novelty of care robots, the lack of current standards guiding their design, and the morally charged contexts within which these robots will be placed, ethical guidelines pertaining to their design and development are not only recommended but are crucial to merit the trust placed in both the designers and the resulting care robot(s).

Secondly, added to the utility of the CCVSD approach is its inherent interdisciplinary focus. Given that care robots draw on multiple disciplines in terms of both their design as well as their use, it is of paramount importance that multiple disciplines be involved in structuring their development and use. But such a task is not an easy feat. To accomplish this, I have translated ethical principles, values and norms into a tangible tool for engineers and robot designers. The CCVSD approach not only allows for interdisciplinary collaboration, but also provides a means to encourage and foster such collaborations. In so doing, the CCVSD approach provides a standardized vocabulary to be used across disciplines, a vocabulary that may be understood by all disciplines involved. Granted the ethicist will have a deeper understanding of ethical principles and the roboticist will have a deeper understanding of robot capabilities, but, all actors in the design process will now have access to a homogenized vocabulary.

Perhaps most notably, the CCVSD approach allows for the ethicist to enter the design process as a member of the design team prior to the robot's development. This is in line with a new wave of ethical technology assessment that seeks to be proactive given the current recognition that technologies are not mere instruments that we use to shape our practices, but rather that they are themselves shaped by society and in turn shape practices and policies by their presence.

In conjunction with the interdisciplinary advantage, a benefit of the CCVSD approach is its commitment to a division of moral labour. The approach encourages an ethical reflection on the part of engineers and designers without demanding that they become specialists; the work of translating ethics has been done for them. Added to this, the approach demands that the ethicist have a technical knowledge of the robot, without demanding they become robot specialists. This is necessary in order to avoid a speculative ethical appraisal of future robots that may never come into existence. The focus is on the creation of robots that may be used currently, based on an understanding of what current robots are capable of.

The CCVSD approach also incorporates the element of implementation where there are additional advantages to draw upon. By proposing a means for the implementation of the care robot, my aim was to marry the intentions of design studies with the results of domestication studies. This would ensure that the built-in morality of the robot would be the same one produced through the use of the robot. By accompanying the technology into its context of use, and engaging in a dialogue with the direct and indirect users, the CCVSD approach once again makes real the care ethics tradition of giving voice to the under-represented (and often under-valued) groups in care; namely, the care-givers. By giving voice to the care-givers and explaining both the initiative and hopes for the care robot, the robot has a better chance of being used in the intended manner. What's more, such a practice is a concrete valuation of the care-givers in their roles. Thus, the CCVSD approach not only bridges the gap between the disciplines involved in the design and the ethical evaluation of the care robot but it also bridges the gap between the intended and actual use of the care robot.

One might also likewise conclude that the insights into the implementation portion might also be used for current care robots which are about to be introduced for the first time in certain hospitals (including surgical robots). In other words, that from this moment on, any robot introduced into a healthcare context is done with the assistance of an ethicist to make clear the re-distribution of roles and responsibilities that will result once the robot is integrated.

In terms of designing future care robots, the work of this book has provided an illustrative example of how to operationalize the VSD approach.

Scholars have criticized VSD for its lack of aligning with any particular ethical theory and have also criticized VSD for being more of a theoretical approach rather than a practical one. To counter the first claim, in aligning VSD with the care ethics tradition this problem was overcome. Added to this, the CCVSD approach provides a pragmatic application of VSD in practice. As an alternative to traditional VSD approaches, I was able to provide a concrete methodology for the design of a future system, one that can be copied for the design of any future care robot rather than being applicable for the design of one specific computer system only.

An additional benefit of the work in this book is seen when we address the CC framework in isolation from the CCVSD approach. In short, the CC framework may be used in the evaluation and future design of other technologies, used in different contexts which are guided by different values. To do this, certain adjustments must be made. If one were to use the framework in the evaluation of an ICT system, one may wish to change the values of importance from the moral elements to values such as privacy, distributive justice, efficiency, or dignity. Accordingly, the practices for evaluation will also adapt to the technology in question. The values chosen for evaluation are dependent on the context of use, the goals of the institution within which it will be used and the end goals of the system. The methodology for describing such practices and the relationship between one practice and another remain the same. Although the values may change, their manner of interpretation (contextualized) does not, nor does the manner in which the practice is described with respect to the interactions between actors as the vehicle for such a manifestation (the methodologies for use).

In closing, I present to you a quote from the work of Joan Tronto: "The best forms of institutional care will be those which are highly deliberate and explicit about how to best meet the needs of the people who they serve" (Tronto, 1993, p. 169). Tronto is mindful of the complexity of care and how demanding such claims are (to understand the care institution and care process in its totality); however, by enforcing three central foci (politics, particularity and purposiveness) the ineffable dimensions of care are uncovered and are articulated. Just as Tronto argues in favour of understanding the care process as a whole in order to structure a good care institution. I too make such a claim with the creation of the CCVSD approach. Including both prospective design as well as implementation through the CCVSD approach I claim that the care process must be understood and critically examined as a whole prior to the design and introduction of a care robot. The manner in which care is understood takes into account the needs of care-receivers but places the needs of care-givers and the allocation of responsibility as central foci.

This line of thinking falls under the umbrella of the structural ethics approach as well, that the care institution be understood in its totality. This demands that the relationship each care practice shares with another be made explicit and criticized: the manifestation of a specific value in one practice impacts the manifestation, or lack thereof, of a specific value or the meeting of needs (of either or both the care-giver and care-receiver) in another practice. To that end, the strong recommendations of Tronto (1993, 2010) and of psychoanalyst Sherry Turkle (2011), coupled with the insights from the structural ethics approach (Brey, 2014), are made real through the CCVSD approach.

In short, the CCVSD approach reflects: a commitment to understanding the starting point of care; being critical of care as it is currently practised; understanding the necessary elements of care and how they contribute to the provision of good care; and above all else how care might change (for better or for worse) with the introduction of a care robot. With these commitments in mind, the future of care, with or without the inclusion of a care robot is studied: the values in care are understood in-depth and preserved. The non-human actors are recognized for their role in forming a morality and meaning within a network. Most importantly, the human care-givers are recognized for the value of their role, and the responsibility of their actions as the stewards of care. Added to this, the CCVSD approach also reflects a commitment to the design and implementation of future care robots that systematically accounts for the valuation of care practices and care workers. Seen through this lens, care robots hold the potential to provide a vital role in aiding the promotion of care values when designed and implemented according to the manner dictated by the CCVSD approach.

I can only hope now that when I enter a hospital and am greeted by a robot, I can rest assured that the robot's role and capabilities have been carefully decided upon to ensure that it adds to my overall good care.

Bibliography

Akrich, M. (1992). The de-scription of technical objects. In W. E. Bijker and J. Law (eds), *Shaping Technology/Building Society: Studies in Sociotechnical Change* (pp. 205–24). Cambridge, MA: MIT Press.

Albrechtslund, A. (2007). Ethics and technology design. *Ethics and Information Technology*, 9 (1), 63–72.

Anderson, M. and Anderson, S. (2007). Machine Ethics: Creating an Ethical Intelligent Agent. *AI Magazine*, 28 (4).

Anderson, M. and Anderson, S. (2010). Robot be good: A call for ethical autonomous machines. *Scientific American*, 303 (4), 15–24.

Argall, B. and Billard, A. (2010). A Survey of Tactile Human-Robot Interactions. *Robotics and Autonomous Systems*, 58 (10), 1159–76.

Arras, K. and Cerqui, D. (2005). *Do we want to share our lives and bodies with robots?* A 2000 People Survey.

Asaro, P. (2006). What should we want from a robot ethic? *International Review of Information Ethics*, 6, 8–16.

Asaro, P. (2009). Modeling the moral user. *Technology and Society Magazine*, IEEE, 28 (1), 20–24.

Barras, C. (2009). Useful, loveable and unbelievably annoying. *The New Scientist*, 22–3.

Beauchamp, T. L. and Childress, J. F. (2001). *Principles of Biomedical Ethics.* Oxford/New York: Oxford University Press.

Beetz, M., Klank, U., Kresse, I., et al. (2011). *Robotic Roommates Making Pancakes.* Humanoid Robots (Humanoids), 2011 11th IEEE-RAS International Conference on.

Bekey, G. (2011). Current Trends in Robotics: Technology and Ethics. In P. Lin, K. Abney and G. Bekey (eds), *Robot Ethics: The Ethical and Social Implications of Robotics* (pp. 17–34). Cambridge, MA: MIT Press.

Bensalem, S., Gallien, M., Ingrand, F., et al. (2009). Designing Autonomous Robots. *IEEE Robotics and Automation Magazine*, 16 (1), 67–77.

Bicchi, A., Peshkin, M. and Colgate, J. (2008). Safety for Physical Human-Robot Interaction. In B. Siciliano and O. Khatib (eds), *Springer Handbook of Robotics* (pp. 1335–48). Berlin: Springer.

Bijker, W. E. and Law, J. (1992). *Shaping Technology/Building Society: Studies in Sociotechnical Change.* Cambridge, MA: MIT Press.

Billard, A., Calinon, S., Dillmann, R. and Schaal, S. (2008). Robot Programming by Demonstration. In B. Siciliano and O. Khatib (eds), *Springer Handbook of Robotics* (pp. 1371–94). Berlin: Springer.

Bischoff, R. and Graefe, V. (2003). HERMES – an Intelligent Humanoid Robot Designed and Tested for Dependability. *Springer Tracts in Advanced Robotics: Experimental Robotics VIII*, 5, 64–74.

Borenstein, J. and Pearson, Y. (2011). Robot Caregivers: Ethical Issues across the Human Life. In P. Lin, K. Abney and G. Bekey (eds), *Robot Ethics: The Ethical and Social Implications of Robotics* (pp. 251–66). Cambridge, MA: MIT Press.

Borgmann, A. (1987). *Technology and the Character of Contemporary Life: A Philosophical Inquiry*. Chicago: University of Chicago Press.

Breazeal, C. L. (2004). *Designing Sociable Robots*. Cambridge, MA: MIT Press.

Breazeal, C. and Aryananda, L. (2002). Recognition of Affective Communicative Intent in Robot-Directed Speech. *Autonomous Robots*, 12 (1), 83–104.

Breazeal, C., Takanishi, A. and Kobayashi, T. (2008). Social Robots that Interact with People. In B. Siciliano and O. Khatib (eds), *Springer Handbook of Robotics* (pp. 1349–70). Berlin: Springer.

Brey, P. (2005). Artifacts as Social Agents. In H. Harbers (ed.), *Inside the Politics of Technology Agency and Normativity in the Co-production of Technology and Society* (pp. 61–84). Amsterdam: Amsterdam University Press.

Brey, P. (2010). Values in Technology and Disclosive Computer Ethics. In L. Floridi (ed.), *The Cambridge Handbook of Information and Computer Ethics* (pp. 41–58). Cambridge: Cambridge University Press.

Brey, P. (2014). From Moral Agents to Moral Factors: The Structural Ethics Approach. In P. Brey (2012). *Anticipatory Ethics for Emerging Technologies. NanoEthics*, 1–13.

Brock, O., Kuffner, J. and Xiao, J. (2008). Motion for Manipulation Tasks. In B. Siciliano and O. Khatib (eds), *Springer Handbook of Robotics* (pp. 615–46). Berlin: Springer.

Buber, M. (1958). *I and Thou*. New York: Scribner.

Buss, M. and Beetz, M. (2010). *CoTeSys – Cognition for Technical Systems*. Kunstliche Intelligenz.

Butterfield, J. (2003). *Collins English Dictionary*. Glasgow: HarperCollins.

Callon, M. (1986). The Sociology of an Actor-Network: The Case of the Electric Vehicle. In M. Callon, J. Law and A. Rip (eds), *Mapping the Dynamics of Science and Technology: Sociology of Science in the Real World* (pp. 19–24). Basingstoke: Macmillan.

Campion, G. and Chung, W. (2008). Wheeled Robots. In B. Siciliano and O. Khatib (eds), *Springer Handbook of Robotics* (pp. 391–410). Berlin: Springer.

Capek, K. and Selver, P. (1923). *R.U.R. (Rossum's Universal Robots): A Fantastic Melodrama*. Garden City, NY: Doubleday, Page & Co.

Capurro, R. (2009). Ethics and Robotics. In R. Capurro and M. Nagenborg (eds), *Ethics and Robotics* (pp. 117–23). Heidelberg; [Amsterdam]: AKA; IOS Press.

Capurro, R. and Nagenborg, M. (2009). *Ethics and Robotics*. Heidelberg; [Amsterdam]: AKA; IOS Press.

Chen, T., King, C.-H., Thomaz, A. and Kemp, C. (2011). *Touched by a Robot: An Investigation of Subjective Responses to Robot-initiated Touch*. Human-Robot Interaction (HRI), 2011 6th ACM/IEEE International Conference on.

Chung, W., Fu, L. and Hsu, S. (2008). Motion Control. In B. Siciliano and O. Khatib (eds), *Springer Handbook of Robotics* (pp. 133–60). Berlin: Springer.

Coeckelbergh, M. (2010). Health care, capabilities, and AI assistive technologies. *Ethical Theory and Moral Practice*, 13 (2), 181–90.

Cooley, M. (2007). From judgment to calculation. *AI & Society*, 21 (4), 395–409.

Cooper, C., Selwood, A., Blanchard, M., et al. (2009). Abuse of people with dementia by family carers: Representative cross sectional survey. *British Medical Journal*, 338 (7694), 583–5.

Correia, M. and Waitzberg, D. (2003). The impact of malnutrition on morbidity, mortality, length of hospital stay and costs evaluated through a multivariate model analysis. *Clinical nutrition (Edinburgh, Scotland)*, 22 (3), 235–9.

Cotin, S., Delingette, H. and Ayache, N. (2000). A hybrid elastic model for real-time cutting, deformations, and force feedback for surgery training and simulation. *VISUAL COMPUTER*, 16, 437–52.

Cummings, M. (2006). Integrating ethics in design through the value-sensitive design approach. *Science and Engineering Ethics*, 12, 701–15.

Daniilidis, K. and Eklundh, J. (2008). 3-D Vision and Recognition. In B. Siciliano and O. Khatib (eds), *Springer Handbook of Robotics* (pp. 543–62). Berlin: Springer.

Dautenhahn, K. (2003). Roles and functions of robots in human society – Implications from research in autism therapy. *Robotica*, 21, 443–52.

Dautenhahn, K. and Werry, I. (2004). Towards interactive robots in autism therapy: Background, motivation and challenges. *Pragmatics & Cognition*, 12 (1), 1–35.

Dautenhahn, K., Woods, S., Kaouri, C., et al. (2005). *What is a Robot Companion – Friend, Assistant or Butler?* Intelligent Robots and Systems, 2005. (IROS 2005). 2005 IEEE/RSJ International Conference on, pp. 1192–7.

Davies, A. and Snaith, P. (1980). Mealtime problems in a continuing-care hospital for the elderly. *Age and Ageing*, 9 (2), 100–105.

Decker, M. (2008). Caregiving robots and ethical reflection: The perspective of interdisciplinary technology assessment. *AI & Society*, 22 (3), 315–30.

Den Hoven, J. van. (2007). *ICT and Value Sensitive Design*. International Federation for Information Processing – Publications – IFIP (233), 67–72.

der, H. van and Reinkensmeyer, D. (2008). Rehabilitation and Health Care Robotics. In B. Siciliano and O. Khatib (eds), *Springer Handbook of Robotics* (pp. 1223–52). Berlin: Springer.

Dietsch, J. (2010). People meeting robots in the workplace. *IEEE Robotics and Automation Magazine*, 17 (2), 15–16.

Dubberly, H. (2004). How do you design? *A Compendium of Models*. http://www.dubberly.com/articles/how-do-you-design.html.

Engelberger, J. F. (1989). *Robotics in Service*. Cambridge, MA: MIT Press.

Evanoff, B. A. (2004). Use of mechanical patient lifts decreased musculoskeletal symptoms and injuries among health care workers. *Injury Prevention*, 10 (4), 212–16.

Facility, T. R. (2008). Our Journey in 2008–9: Annual Report. http://www.torontorehab.com/About-Us/Corporate-Publication/2008-2009/hospital.asp.

Feng, P. and Feenberg, A. (2008). Thinking about Design: Critical Theory of Technology and the Design Process. In P. E. Vermaas (ed.), *Philosophy and Design: From Engineering to Architecture* (pp. 105–18). Dordrecht: Springer.

Feron, E. and Johnson, E. (2008). Aerial Robotics. In B. Siciliano and O. Khatib (eds), *Springer Handbook of Robotics* (pp. 1009–30). Berlin: Springer.

Floridi, L. and Sanders, J. (2004). On the Morality of Artificial Agents. *Minds and Machines*, 14 (3), 349–79.

Fong, T., Nourbakhsh, I. and Dautenhahn, K. (2003). A survey of socially interactive robots. *Robotics and Autonomous Systems*, 42 (3), 143.

Franklin, S. and Graesser, A. (1997). Is it an Agent, or Just a Program? A Taxonomy for Autonomous Agents. *Lecture Notes in Computer Science*, (1193), 21.

Franklin, S., Graesser, A., Olde, B., et al. (1996). *Virtual Mattie – an Intelligent Clerical Agent*. AAA Symposium on Embodied Cognition and Action, Cambridge, MA.

Fraser, N. (1989). *Unruly Practices: Power, Discourse, and Gender in Contemporary Social Theory*. Minneapolis: University of Minnesota Press.

Friedman, B., Kahn, P. H. and Borning, A. (2006). Value Sensitive Design and Information Systems. In P. Zhang and D. Galletta (eds), *Human-Computer Interaction and Management Information Systems: Foundations*. M.E. Sharpe.

Friedrich, H., Münch, S., Dillmann, R., et al. (1996). Robot Programming by Demonstration (RPD): Supporting the Induction by Human Interaction. *Machine Learning*, 23 (2/3), 163–89.

Gadow, S. A. (2002). Nurse and Patient: The Caring Relationship. In A. Bishop and J. Scudder (eds), *Caring, Curing, Coping: Nurse, Physician, and Patient Relationships* (pp. 31–43). Tuscaloosa, AL: University of Alabama Press.

Gill, S. (2008). Socio-ethics of interaction with intelligent interactive technologies. *AI & Society*, 22 (3), 283–300.

Gilligan, C. (1982). *In a Different Voice: Psychological Theory and Women's Development*. Cambridge, MA: Harvard University Press.

Goetz, J. and Kiesler, S. (2002). Cooperation with a Robotic Assistant. CHI Conference on Human Factors in Computing Systems, pp. 578–9.

Hannaford, B. and Okamura, A. (2008). Haptics. In B. Siciliano and O. Khatib (eds), *Springer Handbook of Robotics* (pp. 719–758). Berlin: Springer.

Harbers, H., Mol, A. and Stollmeyer, A. (2002). Food Matters: Arguments for an Ethnography of Daily Care. *Theory, Culture & Society*, 19 (5), 207.

Haselager, W. F. (2005). Robotics, philosophy and the problems of autonomy. *Pragmatics and Cognition*, 13 (3), 515–32.

Hayashi, T. (2005). Control method of robot suit HAL working as operator's muscle using biological and dynamical information. *IEEE*, 3063–8.

Heinzmann, J. and Zelinsky, A. (1998). *3-D Facial Pose and Gaze Point Estimation Using a Robust Real-time Tracking Paradigm*. Automatic Face and Gesture Recognition, 1998. Proceedings. Third IEEE International Conference on, pp. 142–7.

Hertzberg, J. and Chatila, R. (2008). AI Reasoning Methods for Robotics. In B. Siciliano and O. Khatib (eds), *Springer Handbook of Robotics* (pp. 207–28). Berlin: Springer.

Hirano, T. (2007). Generation of Human Care Behaviors by Human-Interactive Robot RI-MAN. *IEEE*, 3128–9.

Hirose, S. and Yamada, H. (2009). Snake-Like Robots. *IEEE Robotics and Automation Magazine*, 16 (1), 88–98.

Hofmann, B. (2008). Why ethics should be part of health technology assessment. *International Journal of Technology Assessment in Health Care*, 24 (4), 423–9.

Hosoda, K., Takuma, T., Nakamoto, A. and Hayashi, S. (2008). Biped robot design powered by antagonistic pneumatic actuators for multi-modal locomotion. *Robotics and Autonomous Systems*, 56 (1), 46–53.

Howcroft, D., Mitev, N. and Wilson, M. (2004). What We May Learn From the Social Shaping of Technology Approach. In J. Mingers and L. Willcocks (eds), *Social Theory and Philosophy for Information Systems* (pp. 329–71). West Sussex, UK: John Wiley and Sons.

Introna, L. (2005). Disclosive Ethics and Information Technology: Disclosing Facial Recognition Systems. *Ethics and Information Technology*, 7 (2), 75–86.

Jain, A. K. and Li, S. Z. (2005). *Handbook of Face Recognition*. New York: Springer Science+Business Media, Inc.

Jecker, N. S. and Self, D. J. (1991). Separating Care and Cure: An Analysis of Historical and Contemporary Images of Nursing and Medicine. *Journal of Medicine and Philosophy*, 16 (3), 285–306.

Jelsma, J. (2006). Designing "moralized" products: Theory and Practice. In P. Verbeek and A. Slob (eds), *User Behavior and Technology Development: Shaping Sustainable Relations between Consumers and Technologies* (pp. 221–31). Dordrecht: Springer.

Kahn, R. E., Swain, M. J., Prokopowicz, P. N. and Firby, R. J. (1996). Gesture Recognition Using the Perseus Architecture. *Computer Vision and Pattern Recognition, IEEE*, 734–41.

Kajita, S. and Espiau, B. (2008). Legged Robots. In B. Siciliano and O. Khatib (eds), *Springer Handbook of Robotics* (pp. 361–90). Berlin: Springer.

Kavraki, L. and LaValle, S. (2008). Motion Planning. In B. Siciliano and O. Khatib (eds), *Springer Handbook of Robotics* (pp. 109–32). Berlin: Springer.

Kawamoto, H. and Sankai, Y. (2002). *Power Assist System HAL-3 for Gait Disorder Person* (pp. 196–203). London, UK: Springer-Verlag.

Kawasaki, H., Komatsu, T., Uchiyama, K. and Kurimoto, T. (1999). *Dexterous Anthropomorphic Robot Hand with Distributed Tactile Sensor: Gifu Hand II.* Mechatronics, IEEE/ASME Transactions on.

Kazerooni, H. (2008). Exoskeletons for Human Performance Augmentation. In B. Siciliano and O. Khatib (eds), *Springer Handbook of Robotics* (pp. 773–98). Berlin: Springer.

Kemp, C., Fitzpatrick, P., Hirukawa, H., et al. (2008). Humanoids. In B. Siciliano and O. Khatib (eds), *Springer Handbook of Robotics* (pp. 1307–34). Berlin: Springer.

Kidd, C. D. (2008). *Designing for Long-term Human–Robot Interaction and Application to Weight Loss.* PhD dissertation.

Kidd, C. D. and Breazeal, C. L. (2006). *Designing a Sociable Robot System for Weight Maintenance, IEEE*, 253–7.

Kim, K. H., Bang, S. W. and Kim, S. R. (2004). Emotion recognition system using short-term monitoring of physiological signals. *Medical & Biological Engineering & Computing*, 42, 419–27.

Kiran, A. H. (2011). Responsible Design. A Conceptual Look at Interdependent Design-Use Dynamics. *Philosophy & Technology* (1).

Koggel, C. M. (1998). *Perspectives on Equality: Constructing a Relational Theory.* Lanham, MD: Rowman & Littlefield Publishers.

Koughnett, J. V., Jayaraman, S., Eagleson, R., et al. (2009). Are there advantages to robotic-assisted surgery over laparoscopy from the surgeon's perspective? *Journal of Robotic Surgery*, 3, 79–82.

Krapp, K. (2002). Activities of Daily Living Evaluation. In *Encyclopedia of Nursing & Allied Health.* Detroit, MI: Gale Group, Inc.

Kubrick, S., Clarke, A. C., Dullea, K., et al. (1968). *2001, A Space Odyssey.* Burbank, CA: Warner Home Video.

Kunze, L., Roehm, T., Beetz, M. (2011). *Towards Semantic Robot Description Languages.* Robotics and Automation (ICRA), 2011 IEEE International Conference on, pp. 5589–95.

Latour, B. (1992). Where Are the Missing Masses? The Sociology of a Few Mundane Artifacts. In W. Bijker and J. Law (eds), *Shaping Technology/Building Society: Studies in Sociotechnical Change* (pp. 225–58). Cambridge, MA: MIT Press.

Lauwers, T. B., Kantor, G. A. and Hollis, R. L. (2006). *A Dynamically Stable Single-wheeled Mobile Robot with Inverse Mouse-ball Drive.* Robotics and Automation, 2006. ICRA 2006. Proceedings 2006 IEEE International Conference on, pp. 2884–9.

Le, C. A., Poole, E. S. and Wyche, S. P. (2009). *Values as Lived Experience: Evolving Value Sensitive Design in Support of Value Discovery* (pp. 1141–50). New York: ACM.

Leininger, M. (1988). Leininger's theory of nursing: Culture care diversity and universality. *Nursing Science Quarterly*, 2, 11–20.

Lin, P., Abney, K. and Bekey, G. (2011). *Robot Ethics: The Ethical and Social Implications of Robotics.* Cambridge, MA: MIT Press.

Little, M. O. (1998). Care: From Theory to Orientation and Back. *The Journal of Medicine and Philosophy*, 23 (2), 190–209.

Lofquist, L. H. and Dawis, R. (1978). Values as second-order needs in the theory of work adjustment. *Journal of Vocational Behavior*, 12 (1), 12–19.

Lytle, M. (2002). Robot care bears for the elderly. *BBC.* http://news.bbc.co.uk/2/hi/science/nature/1829021.stm.

MacDorman, K. F. and Ishiguro, H. (2006). The uncanny advantage of using androids in cognitive and social science research. *Interaction Studies*, 7 (3), 297–337.

Manders-Huits, N. (2011). What Values in Design? The Challenge of Incorporating Moral Values into Design. *Science and Engineering Ethics*, 17 (2), 271–87.

Maslow, A. H. (1970). *Motivation and Personality.* New York: Harper & Row.

Maurer, M. (2007). *Some Ideas on ICT as it Influences the Future.* NEC Technology Forum, Tokyo.

Max-Neef, M. (1995). Economic growth and quality of life: A threshold hypothesis. *Ecological Economics*, 15 (2), 115–18.

Mayer, C., Radig, B., Sosnowski, S. and Kuhnlenz, K. (2010). *Towards Robotic Facial Mimicry: System Development and Evaluation.* Proceedings – IEEE International Workshop on Robot and Human Interactive Communication, pp. 198–203.

Melchiorri, C. and Kaneko, M. (2008). Robot Hands. In B. Siciliano and O. Khatib (eds), *Springer Handbook of Robotics* (pp. 345–360). Berlin: Springer.

Metzler, T. and Lewis, L. (2008). *Ethical Views, Religious Views, and Acceptance of Robotic Applications: A Pilot Study.* Association for the Advancement of Artificial Intelligence (www.aaai.org), 15–22.

Minato, T., Shimada, M., Ishiguro, H. and Itakura, S. (2004). Development of an Android Robot for Studying Human-Robot Interaction. *Lecture Notes in Computer Science* (3029), 424–34.

Minato, T., Yoshikawa, Y., Noda, T., et al. (2007). *CB2: A Child Robot with Biomimetic Body for Cognitive Developmental Robotics.* Humanoid Robots, 2007 7th IEEE-RAS International Conference on, pp. 557–62.

Minguez, J., Lamiraux, F. and Laumond, J. (2008). Motion Planning and Obstacle Avoidance. In B. Siciliano and O. Khatib (eds), *Springer Handbook of Robotics* (pp. 827–852). Berlin: Springer.

Mitcham, C. (2005). *Encyclopedia of Science, Technology and Ethics*. Detroit, MI: Macmillan Reference.

Mitra, P. and Niemeyer, G. (2008). Model-mediated Telemanipulation. *The International Journal of Robotics Research*, 27 (2), 253–62.

Mol, A. (2010). Care and its values: Good food in the nursing home. In A. Mol, I. Moser and A. Pols (eds), *Care in Practice: On Tinkering in Clinics, Homes and Farms*. Bielefeld: Transcript Verlag.

Mol, A., Moser, I. and Pols, J. (2010). *Care in Practice: On Tinkering in Clinics, Homes and Farms*. Bielefeld; Piscataway, NJ: Transcript; Distributed in North America by Transaction Publishers.

Moor, J. H. (1995). Is Ethics Computable? *Metaphilosophy*, 26 (1–2), 1.

Moor, J. H. (2006). Machine Ethics – The Nature, Importance, and Difficulty of Machine Ethics. *IEEE Intelligent Systems*, 21 (4), 18.

Moravec, H. P. (1999). *Robot: Mere Machine to Transcendent Mind*. New York: Oxford University Press.

Mori, M. (1970). Bukimi no tani: The uncanny valley. *Energy*, 7 (4), 33–5.

Morin, P. and Samson, C. (2008). Motion Control of Wheeled Mobile Robots. In B. Siciliano and O. Khatib (eds), *Springer Handbook of Robotics* (pp. 799–826). Berlin: Springer.

Mowshowitz, A. (2008). Technology as excuse for questionable ethics. *AI & Society*, 22 (3), 271–82.

Mutlu, B. and Forlizzi, J. (2008). *Robots in Organizations: The Role of Workflow, Social, and Environmental Factors in Human-robot Interaction*. Human-Robot Interaction (HRI), 2008, 3rd ACM/IEEE International Conference, pp. 287–94.

Nathan, L. P., Friedman, B., Klasnja, P., et al. (2008). *Envisioning Systemic Effects on Persons and Society throughout Interactive System Design* (pp. 1–10). USA: ACM.

Nedelsky, J. (2008). Reconceiving Rights and Constitutionalism. *Journal of Human Rights*, 7 (2), 139–73.

Neven, L. (2010). "But obviously not for me": Robots, laboratories and the defiant identity of elder test users. *Sociology of Health and Illness*, 32 (2), 335–47.

Niemeyer, G., Preusche, C. and Hirzinger, G. (2008). Telerobotics. In B. Siciliano and O. Khatib (eds), *Springer Handbook of Robotics* (pp. 741–58). Berlin: Springer.

Nissenbaum, H. (2001). How computer systems embody values. *Computer*, 34 (3), 120–19.

Noddings, N. (1984). *Caring, a Feminine Approach to Ethics & Moral Education*. Berkeley: University of California Press.

Noddings, N. (2002). *Starting at Home Caring and Social Policy*. Berkeley: University of California Press.

Nordmann, A. and Rip, A. (2009). Mind the gap revisited. *Nature Nanotechnology*, 4 (5), 273–4.

Nurses, College of (1999). *Standard for the Therapeutic Nurse-client Relationship: For Registered Nurses and Registered Practical Nurses in Ontario: Standards of Practice*. The College of Nurses of Ontario.

Nurses, College of (1999). *The Ethical Framework for Nurses in Ontario: Standards of Practice*. The College of Nurses of Ontario.

Nussbaum, M. C. (2000). *Women and Human Development: The Capabilities Approach*. Cambridge/New York: Cambridge University Press.

Oh, J.-H., Hanson, D., Kim, W.-S., et al. (2006). *Design of Android Type Humanoid Robot Albert HUBO*. Intelligent Robots and Systems, 2006 IEEE/RSJ International Conference on, pp. 1428–33.

Oosterhof, N. (2005). *Thinking Machines that Feel: The Role of Emotions in Artificial Intelligence Research*. Master's thesis, University of Twente.

Oosterlaken, I. (2009). Design for development: A capability approach. *Design Issues*, 25 (4), 91–102.

Orpwood, R., Adlam, T., Evans, N. and Chadd, J. (2008). Evaluation of an assisted-living smart home for someone with dementia. *Journal of Assistive Technologies*, 2 (2), 13–21.

Payne, B. K. and Cikovic, R. (1995). Empirical Examination of the Characteristics, Consequences, and Causes of Elder Abuse in Nursing Homes. *Journal of Elder Abuse & Neglect*, 7 (4), 61–74.

Pellegrino, E. D. (1985). The Virtuous Physician, and the Ethics of Medicine. In E. Shelp (ed.), *Virtue and Medicine: Explorations in the Character of Medicine* (Vol. 1). Dordrecht: D. Reidel Publishing Company.

Picard, R. W. (2000). *Affective Computing*. Cambridge, MA: MIT Press.

Pillemer, K. and Moore, D. W. (1990). Highlights from a Study of Abuse of Patients in Nursing Homes. *Journal of Elder Abuse & Neglect*, 2 (1–2), 5–29.

Pineau, J., Montemerlo, M., Pollack, M., et al. (2003). Towards robotic assistants in nursing homes: Challenges and results. *Robotics and Autonomous Systems*, 42 (3), 271.

Podnieks, E. (1990). *National Survey on Abuse of the Elderly in Canada*. Toronto, Ontario: Ryerson Polytechnical Institute.

Pollack, M. E., Brown, L., Colbry, D., et al. (2002). *Pearl: A mobile robotic assistant for the elderly*. AAAI 2002, Workshop on Automation as Caregiver: The Role of Intelligent Technology in Elder Care, pp. 85–92.

Pols, A. J. (2004). *Good Care: Enacting a Complex Ideal in Long-term Psychiatry*. Utrecht: Trimbos-instituut.

Prattichizzo, D. and Trinkle, J. (2008). Grasping. In B. Siciliano and O. Khatib (eds), *Springer Handbook of Robotics* (pp. 671–700). Berlin: Springer.

Rayman, R., Croome, K., Galbraith, N., et al. (2006). Long-distance robotic telesurgery: A feasibility study for care in remote environments. *The International Journal of Medical Robotics + Computer Assisted Surgery*, 2 (3), 216–24.

Rayman, R., Croome, K., Galbraith, N., et al. (2007). Robotic telesurgery: A real-world comparison of ground- and satellite-based internet performance. *The International Journal of Medical Robotics + Computer Assisted Surgery*, 3 (2), 111–16.

Razavi, S. (2007). *The Political and Social Economy of Care in a Development Context: Conceptual Issues, Research Questions and Policy Options*. Geneva: United Nations Research Institute for Social Development.

Reich, W. T. (1995). History of the notion of care. In W. T. Reich (ed.), *Encyclopedia of Bioethics* (pp. 219–331). New York/London: Macmillan; Simon & Schuster; Prentice Hall International.

Roach, M. S. (1999). *The Human Act of Caring: A Blueprint for the Health Professions*. Ottawa: Canadian Hospital Association Press.

Robotics, A. (2010). Nao, the ideal partner for research and robotics classrooms. www.aldebaran-robotics.com/en.

Ruddick, S. (1995). *Maternal Thinking: Toward a Politics of Peace* [with a new preface]. Boston: Beacon Press.

Rudnick, A. (2001). A meta-ethical critique of care ethics. *Theoretical Medicine and Bioethics*, 22 (6), 505–17.

Russell, S. J. and Norvig, P. (1995). *Artificial Intelligence: A Modern Approach*. Englewood Cliffs, NJ: Prentice Hall.

Saenz, A. (2010). Incredible TUG Robots Automate Delivery in Hospitals. Singularity Hub. http://singularityhub.com/2010/06/06/incredible-tug-robots-automate-delivery-in-hospitals-video/.

Sakamoto, D., Kanda, T., Ono, T., et al. (2007). *Android as a Telecommunication Medium with a Human-like Presence* (pp. 193–200). USA: ACM.

Sandelowski, M. (1997). Exploring the gender-technology relation in nursing. *Nursing Inquiry*, 4 (4), 219–28.

Santoro, M., Marino, D. and Tamburrini, G. (2008). Learning robots interacting with humans: From epistemic risk to responsibility. *AI & Society*, 22 (3), 301–14.

Satoh, H., Kawabata, T. and Sankai, Y. (2009). *Bathing Care Assistance with Robot Suit HAL*, IEEE, pp. 498–503.

Schoenhofer, S. (2001). A Framework for Caring in a Technologically Dependent Nursing Practice Environment. In *Advancing Technology, Caring and Nursing* (Rozzano Loscin ed., pp. 3–11). Westport, CT: Auburn House.

Schulz, D., Burgard, W., Fox, D. and Cremers, A. B. (2003). People Tracking with Mobile Robots Using Sample-Based Joint Probabilistic Data Association Filters. *The International Journal of Robotics Research*, 22 (2), 99–116.

Sen, A. (1985). *Commodities and Capabilities*. Amsterdam/New York: Elsevier.

Sharkey, N. and Sharkey, A. (2011). The Rights and Wrongs of Robot Care. In P. Lin, K. Abney and G. Bekey (eds), *Robot Ethics: The Ethical and Social Implications of Robotics* (pp. 267–82). Cambridge, MA: MIT Press.

Sharkey, A. and Sharkey, N. (2012). Granny and the robots: Ethical issues in robot care for the elderly. *Ethics and Information Technology*, 14 (1), 27–40.

Shaw-Garlock, G. (2009). Looking Forward to Sociable Robots. *International Journal of Social Robotics*, 1 (3), 249–60.

Shieh, M. Y., Lu, C. M., Chen, C. C., et al. (2007). *Design and Implementation of an Interactive Nurse Robot.* SICE, 2007 Annual Conference, pp. 2121–5.

Siciliano, B. and Khatib, O. (2008). *Springer Handbook of Robotics*. Berlin: Springer.

Sidner, C. L. and Dzikovska, M. (2005). A First Experiment in Engagement for Human-Robot Interaction in Hosting Activities. In J. C. J. van Kuppevelt, L. Dybkjær, N. O. Bernsen and N. Ide (eds), *Advances in Natural Multimodal Dialogue Systems*, Vol. 30 (pp. 55–76). Netherlands: Springer.

Silverstone, R., Hirsch, E. and Morley, D. (1992). Information and communication technologies and the moral economy of the household. In R. Silverstone and E. Hirsch (eds), *Consuming Technologies: Media and Information in Domestic Spaces* (pp. 15–31). London: Routledge.

Singer, P. W. (2009). *Wired for War: The Robotics Revolution and Conflict in the Twenty-first Century*. New York: Penguin Press.

Smits, R., Leyten, J. and den Hertog, P. (1995). Technology assessment and technology policy in Europe: New concepts, new goals, new infrastructures. *Policy Sciences*, 28 (3), 271–99.

Soraker, J. (forthcoming). When Does the Mind matter? The Strengths and Limitations of the Informational Level of Abstraction. *Ethics and Information Technology*.

Sorensen, K. (2005). Domestication: The Enactment of Technology. In T. Berker, M. Hartmann, Y. Punie and K. Ward (ed.), *Domestication of Media and Technology* (pp. 40–61). Open University Press.

Sparrow, R. and Sparrow, L. (2006). In the hands of machines? The future of aged care. *Minds and Machines*, 16 (2), 141–61.

Stabell, A., Eide, H., Solheim, G. A., et al. (2004). Nursing Older People: Nursing home residents' dependence and independence. *Journal of Clinical Nursing*, 13 (6), 677–86.

Sullins, J. (2006) When is a robot a moral agent? *International Review of Information Ethics*, 6, 23–30.

Super, D. E. (1968). *Work Values Inventory*. Boston: Houghton Mifflin.

Sutton, R. S. and Barto, A. G. (2010). *Reinforcement learning: An introduction.* Cambridge, MA: MIT Press.

Swierstra, T. and Rip, A. (2007). Nano-ethics as NEST-ethics: Patterns of Moral Argumentation about New and Emerging Science and Technology. *NanoEthics*, 1 (1), 3–20.

Tamburrini, G. (2009). Robot Ethics: A View from the Philosophy of Science. In R. Capurro and M. Nagenborg (eds), *Ethics and Robotics* (pp. 11–22). Heidelberg; [Amsterdam]: AKA; IOS Press.

Tamura, T., Yonemitsu, S., Itoh, A., et al. (2004). Is an entertainment robot useful in the care of elderly people with severe dementia? *The Journals of Gerontology Series A: Biological Sciences and Medical Sciences*, 59 (1), M83–M85.

Tenorth, M. (2011). *Knowledge Processing for Autonomous Robots*. PhD dissertation, München: Universitätsbibliothek der TU München.

Tenorth, M. and Beetz, M. (2012). Knowledge Processing for Autonomous Robot Control. In *AAAI Spring Symposium: Designing Intelligent Robots*.

Tenorth, M., Jain, D. and Beetz, M. (2010). Knowledge Representation for Cognitive Robots. *Kunstliche Intelligenz*, 24 (3), 233–40.

Thaler, R. H. and Sunstein, C. R. (2008). *Nudge: Improving Decisions about Health, Wealth, and Happiness*. New Haven: Yale University Press.

Thow-Hing, V. N., Torisson, K., Sarvadevabhatla, R. K., et al. (2009). Cognitive Map Architecture: Facilitation of Human-Robot Interaction in Humanoid Robots. *IEEE Robotics and Automation Magazine*, 16 (1), 55–66.

Thrun, S. (2004). Toward a Framework for Human-Robot Interaction. *Human-Computer Interaction*, 19 (1), 9–24.

Thrun, S., Schulte, J. and Rosenberg, C. (2000). Interaction with Mobile Robots in Public Places. *IEEE Intelligent Systems*, 7–11.

Torrance, S. (2008). Ethics and consciousness in artificial agents. *AI & Society*, 22 (4), 495–521.

Tronto, J. (2010). Creating Caring Institutions: Politics, Plurality, and Purpose. *Ethics and Social Welfare*, 4 (2), 158–71.

Tronto, J. C. (1993). *Moral Boundaries: A Political Argument for an Ethic of Care*. New York: Routledge.

Turing, A. M. (1950). Computing machinery and intelligence. *Mind: A Quarterly Review of Psychology and Philosophy*, LIX (236), 433.

Turkle, S. (2011). *Alone Together: Why We Expect More from Technology and Less from Each Other*. New York: Basic Books.

Vallor, S. (2011). Carebots and caregivers: Sustaining the ethical ideal of care in the twenty-first century. *Philosophy and Technology*, 24 (3), 251–68.

van de Poel, I. (2009). Values in engineering design. In A. Meijers (ed.), *Handbook of the Philosophy of Science. Volume 9: Philosophy of Technology and Engineering Sciences*. Oxford: Elsevier.

van de Poel, I. and Kroes, P. (2014). Can Technology Embody Values? In P. Kroes and P.-P. Verbeek (eds), *Moral Agency and Technical Artefacts*. Dordrecht: Springer.

van der Plas, A., Smits, M. and Wehrmann, C. (2010). Beyond speculative robot ethics: A vision assessment study on the future of the robotic caretaker. *Accountability in Research*, 17, 6, 299–315.

van Gorp, A. And Van de Poel, I. (2008). Deciding on Ethical Issues in Engineering Design. In P. E. Vermaas (ed.), *Philosophy and Design: From Engineering to Architecture* (pp. 77–90). Dordrecht: Springer.

van Wynsberghe, A. (2013a). Designing robots for care: Care centered value-sensitive design. *Science and Engineering Ethics*, 19(2), 407–33.

van Wynsberghe, A. (2013b). A Method for Integrating Ethics into the Design of Robots. *Industrial Robot*, 40(5), 433–40.

van Wynsberghe, A. (2014). *To Delegate or not to Delegate: Care Robots, Moral Agency and Moral Responsibility*. Machine Ethics in the Context of Medical and Care Agents, conference proceedings, April.

van Wynsberghe, A. and Gastmans, C. (2008). Telesurgery: An ethical appraisal. *Journal of Medical Ethics*, 34 (10).

van Wynsberghe, A. and Gastmans, C. (2009). Telepsychiatry and the meaning of in-person contact: A preliminary ethical appraisal. *Medicine, Health Care, and Philosophy*, 12 (4), 469–76.

van Wynsberghe, A. and Robbins, S. (2014). Ethicist as Designer: A pragmatic approach to ethics in the lab. *Science and Engineering Ethics*, 20(4), 947–61.

Vanlaere, L. and Gastmans, C. (2011). A personalist approach to care ethics. *Nursing Ethics*, 18 (2), 161–73.

Verbeek, P. (2008). Morality in Design; design ethics and the morality of technological artifacts. In P. E. Vermaas (ed.), *Philosophy and Design: From Engineering to Architecture* (pp. 91–102). Dordrecht: Springer.

Verbeek, P.-P. (2005). *What things Do: Philosophical Reflections on Technology, Agency, and Design*. Pennsylvania.: Pennsylvania State University Press.

Verbeek, P.-P. (2006). Materializing Morality. *Science, Technology, & Human Values*, 31 (3), 361–80.

Verbeek, P.-P. (2011). *Moralizing Technology: Understanding and Designing the Morality of Things*. Chicago/London: The University of Chicago Press.

Verkerk, M. (2001). The care perspective and autonomy. *Medicine, Health Care, and Philosophy*, 4 (3), 289–94.

Verkerk, M. A., Busschbach, J. J. V. and Karssing, E. D. (2001). Health-Related Quality of Life Research and the Capability Approach of Amartya Sen. *Quality of Life Research*, 10 (1), 49–55.

Veruggio, G. and Abney, K. (2011). Roboethics: The Applied Ethics for a New Science. In P. Lin, K. Abney and G. Bekey (eds), *Robot Ethics: The Ethical and Social Implications of Robotics* (pp. 347–64). Cambridge, MA: MIT Press.

Veruggio, G. and Operto, F. (2006). Roboethics: A Bottom-up Interdisciplinary Discourse in the Field of Applied Ethics in Robotics. *International Review of Information Ethics*, 6, 3–8.

Veruggio, G. and Operto, F. (2008). Roboethics: Social and Ethical Implications of Robotics. In B. Siciliano and O. Khatib (eds), *Springer Handbook of Robotics* (pp. 1499–524). Berlin: Springer.

Villani, L. and de Schutter, J. (2008). Force Control. In B. Siciliano and O. Khatib (eds), *Springer Handbook of Robotics* (pp. 161–86). Berlin: Springer.

Vongsoasup, V. and Mataric, M. (n.d.). *Path Planning and Navigation for Geeves: A Tour-Guide/Greeter Robot.*

Wada, K., Shibata, T., Saito, T., et al. (2005). *Psychological and Social Effects of One Year Robot Assisted Activity on Elderly People at a Health Service Facility for the Aged.* Robotics and Automation, 2005. ICRA 2005. Proceedings of the 2005 IEEE International Conference on, pp. 2785–90.

WALL-E [Motion picture]. (2008). Walt Disney Home Entertainment.

Wallach, W. (2010). Robot minds and human ethics: The need for a comprehensive model of moral decision making. *Ethics and Information Technology*, 12 (3), 243–50.

Wallach, W. and Allen, C. (2010). *Moral Machines: Teaching Robots Right from Wrong.* New York/Oxford: Oxford University Press.

Wallach, W., Allen, C. and Smit, I. (2008). Machine morality: Bottom-up and top-down approaches for modelling human moral faculties. *AI & Society*, 22 (4), 565–82.

Wallach, W., Franklin, S. and Allen, C. (2010). A conceptual and computational model of moral decision making in human and artificial agents. *Topics in Cognitive Science*, 2 (3), 454–85.

Walters, M. L., Dautenhahn, K., Woods, S. N. and Koay, K. L. (2007). *Robotic Etiquette: Results from User Studies Involving a Fetch and Carry Task.* Human-Robot Interaction (HRI), 2007 2nd ACM/IEEE International Conference on, pp. 317–24.

Weaver, K., Morse, J. and Mitcham, C. (2008). Ethical sensitivity in professional practice: Concept analysis. *Journal of Advanced Nursing*, 62 (5), 607–18.

Widdershoven, G. (2002). Technology and Care, from Opposition to Integration. In C. Gastmans (ed.), *Between Technology and Humanity: The Impact of Technology on Health Care Ethics* (pp. 35–48). Leuven: Leuven University Press.

Wilson, M. (2002). Making nursing visible? Gender, technology and the care plan as script. *Information Technology & People*, 15 (2), 139–58.

Wong, P.-H. (2011). Technology, Recommendation and Design: On Being a 'Paternalistic' Philosopher. *Science and Engineering Ethics*, 19 (1), 27–42.

Wood, R. (2008). Fly, robot fly. *IEEE Spectrum*, 45 (3), 25–9.

Wright, L., Hickson, M. and Frost, G. (2006). Eating together is important: Using a dining room in an acute elderly medical ward increases energy intake. *Journal of Human Nutrition and Dietetics*, 19 (1), 23–26.

Yoshiro, U., Shinichi, O., Yosuke, T., et al. (2005). *Childcare Robot PaPeRo is Designed to Play with and Watch over Children at Nursery, Kindergarten, School and at Home.* Development of Childcare Robot PaPeRo, 1–11.

Zaeh, M. F., Roesel, W., Bannat, A., et al. (2010). Artificial Cognition in Production Systems. *IEEE Transactions on Automation Science and Engineering*, 7 (3), 1–27.

Zhang, T., Zhu, B., Lee, L. and Kaber, D. (2008). *Service Robot Anthropomorphism and Interface Design for Emotion in Human–Robot Interaction*, IEEE, pp. 674–9.

Index